Linear Audo - Volume 7

AF879424

Jan Didden
publisher/editor

Authors:

Stan Curtis

Jan Didden

Joachim Gerhard

Marcel van de Gevel

Barrie Gilbert

Malcolm Hawksford

George Ntanavaras

Douglas Self

Giovanni Stochino

Burkhard Vogel

Stuart Yaniger

David Zan

Linear Audio is published by Linear Audio of Turnhout, Belgium.
Jan Didden, Publisher/Managing Editor
Vol 7 is published 1 March 2014.

Important notices and disclaimers

Many of the designs, projects, products, methods, instructions or ideas in Linear Audio publications involve potentially lethal voltages. The information is provided with the understanding that the practitioner has sufficient knowledge and experience to safely execute the ideas and practices described in this publication, and to avoid injury and damages to persons and property. Safe and professional practices are assumed. If you are not sure to have this knowledge and experience, you should seek experienced professional advice and assistance.

Neither the Publisher nor the authors, contributors or editors assume any liability for any injury and/or damage to persons or property as a matter of product liability, negligence or otherwise, or from any use or operation of any projects, products, methods or ideas contained in the material in Linear Audio publications and materials.

The designs, projects, products, methods, instructions or ideas in Linear Audio publications may be subject to patents, trademarks or similar deposits, which are all recognized.

The use of general descriptive or registered names, trademarks or similar identifications in this publication does not imply that these are free for general use.

Design, Prepress: Fred Zurel, F&N Eigen Beheer - www.eigenbeheer.nl
Production: F&N Eigen Beheer, Castricum, The Netherlands

ISBN 978-949092-9084

www.linearaudio.net

Contents

Letters to the Editor: I encourage readers' comments on articles in Linear Audio, and I regularly receive them. I will post these, together with the author's reply, on-line at www.linearaudio.net

The G-Pole – An Adventure in Space[1]

Joachim Gerhard

When did you have your last enlightenment?
Maybe as a child. Many. Sugar is sweet, fire is hot. Then your first pill: bitter.
Did you even have Satori? An intense feeling that you can love EVERYTHING. Stones, flowers, every human being! I must admit it was not that intense but I was quite mesmerized when it struck me that there is something wrong with conventional dipoles.
It all started back in the same year as 9/11.
There was a Stereophile show in May in New York. I even had a room in the same Marriot hotel that fell to pieces some months later.
I walked into a room where a strongly beaming electrostatic speaker was playing. At first I was told to sit in the sweet spot to enjoy the soundstage but I did not get that far.

One meter into the room and I heard the sound loud and clear radiating from two points in space on the back wall totally detached from the physical speakers. By slowly walking around and finding that "virtual" spot where the reflected signals coincided I could even hear a "second" stereo image. Several months later I had nearly forgotten that event but memory came back even stronger when I visited the son of the late and great Harold Beveridge, Rick, in Sonoma County. In a major effort, Rick had put together everything he could find about the audio work of his father and had set up shop to produce improved versions of the vintage and legendary speakers. We talked about his father's philosophy and especially his thoughts about playing WITH the room and not AGAINST it. His solution was a line source with a lens. It had electrostatic elements, damped on the back and an acoustic lens in the front. Those speakers where set up on the side walls, radiating *towards* each other. They could throw an extremely wide and deep soundstage. When you walked between the speakers to the other side of the room you could hear the soundstage again. Left and right reversed!

Suddenly a lot of things fell into place. I made a small drawing for Rick (**figure 1**). Two strongly beaming electrostatics could be set up in such a way that you listen on the cancelation axis. One image is

[1] Editor's note: This Guest Editorial is based on Joachim Gerhard's presentation at the 2013 European Triode Festival, held in Berlin in October 2013. It is not a traditional technical article but documents his discovery trip of many years into sound reproduction and acoustics. I am grateful for his willingness to share this very personal adventure with us.

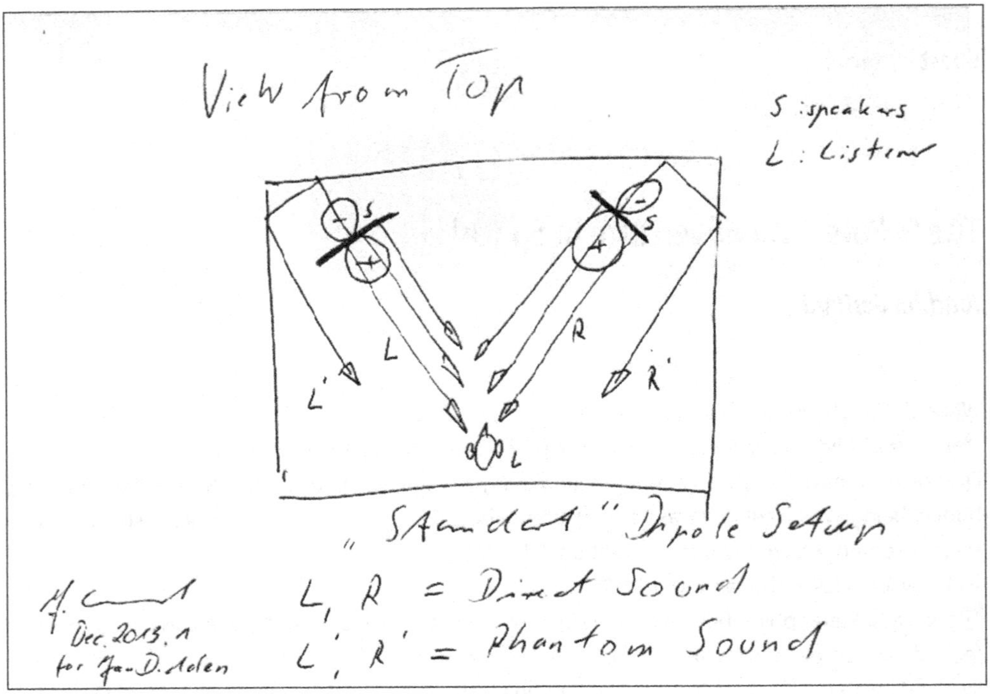

Figure 1

thrown against the front wall and one image is thrown against the sidewall, reaching the ears with considerable delay. Sitting in the middle you should hear a well-developed soundstage and the delayed sound from the side walls could enhance spaciousness without blurring the image. Rick got the essence immediately and told me that this is the first good idea how to drive a room since his father had passed. Yes, it's true, ask him. You can imagine how proud I was.

Back home I had another idea. The beaming positive side that reflects from the front wall could be thrown against a curved plate. That would also solve the beaming problem making it possible to hear with several people (**figure 2**). Now I was hooked and went into serious research.

I set up a wide band speaker on an open baffle and started to measure the first strong reflection. I even was granted a patent "Gekreutze Schallwand" (Crossed Sound Surface) where two baffles were arranged in such a way that the woofer section can be rotated horizontally against the midrange section. There is not enough space here to describe it in full. Read the patent if you wish. The beaming electrostatic experience has survived here in the form of a "Phantom Source", so this is the first patent I know of that claims that a stereo phantom source can be used to listen to high end audio. My curiosity did not stop there so I started to work with several baffles in the same room. I think I ended up with 6 baffles. I fed them the same mono signal and after some experimentation was able

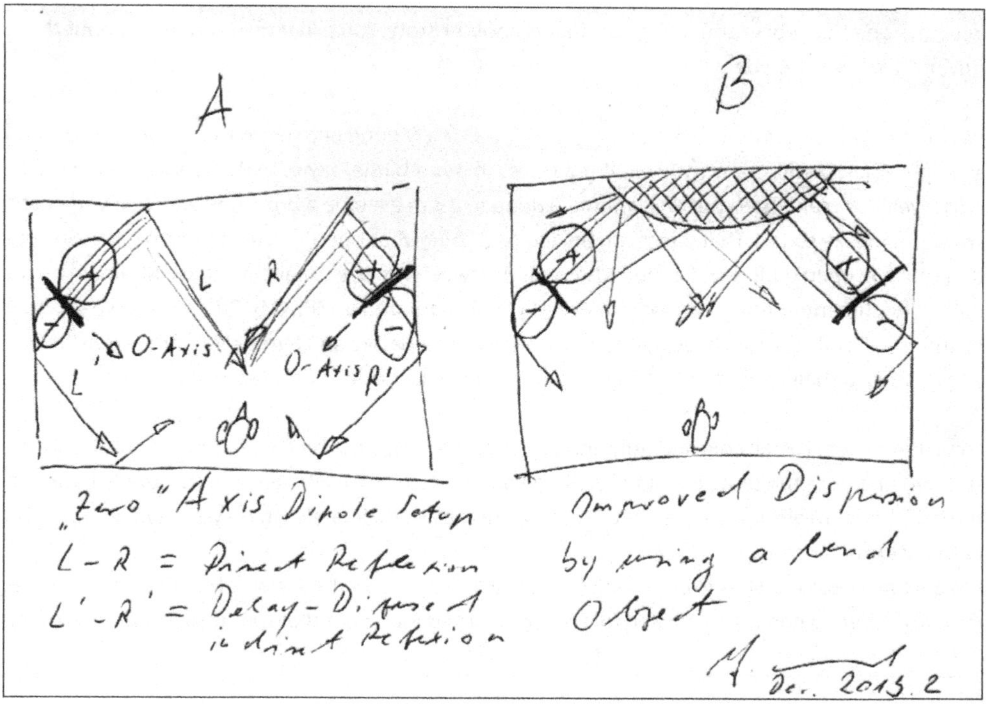

Figure 2

to throw something like a fake stereo signal. People where really surprised to hear a setup like that. I called the effect "Quellenseparation" that would translate into something like "Source Separation". I thought it would be great for artists that work with sound or such but ultimately it drove me nuts. Too much work for a single person. So for some time I stopped this work and constructed more or less conventional speakers, phono stages and such.

My interest in open baffles came back when a friend of mine developed a new wide-band speaker. Actually he tried to combine the best of the old and the new and he succeeded big time after more than two years of experimentation. The most difficult part was to find a membrane that does not break up too much. Instead of following the usual way of using at least one coincident frequency where the membrane breaks up and start to work as a bending wave radiator, not much different from the Manger Transducer, that has simply a very low coincident frequency, he was searching for a membrane that is as stiff as possible, even at higher frequencies. He also wanted to avoid a whizzer cone because that is again a new coincident frequency higher up. His reasoning is that bending wave transducers can create a lot of harmonic distortion, especially second. That is always a sign of chaos and can lead to masking of information although it can be quite pleasing to the ear. The whizzer cone on many drivers also works in opposite phase to the cone. You can see that as a hole in the response around 10kHz in many drivers. Years later my friend was even able to make a version with a whizzer cone that works admirably, but that is another story. You can find information about those drivers at www.enviee.de .

Back to business. I was convinced that this driver was really good and found that it sounded best on a simple open baffle. Try it yourself. Put a driver in a big baffle, say 0.2 x 1.2m and listen without crossover, just plain. Then put any volume behind it, it can even be a box from lead, and you hear a strange squeaky sound. That effect can be damped somewhat but the virgin magic is gone. So I put that driver in open baffles and added a horn ribbon tweeter over 8kHz and an active M-dipole below 80Hz. I documented much of that work on diyaudio.com under MPL (My Private Loudspeaker). It sounds swell and I got much recognition, also from Holger Barske and Michael Methe who visited me and described their experience on the web. (Holger Barske.de Michael Methe.de).

Well, that system is really big and ugly although I plan to make the baffles somewhat prettier. And it suffered first from the same effect I described earlier. When you walk between the dipole baffles to the back and turn around you can hear a second image approximately where you have being sitting before (**figure 3**).

For a while I kept it just as is but then something struck me on the www.linkwitzlab.com website. Actually Siegfried had called me and told me that he had found out that the latest incarnation of the

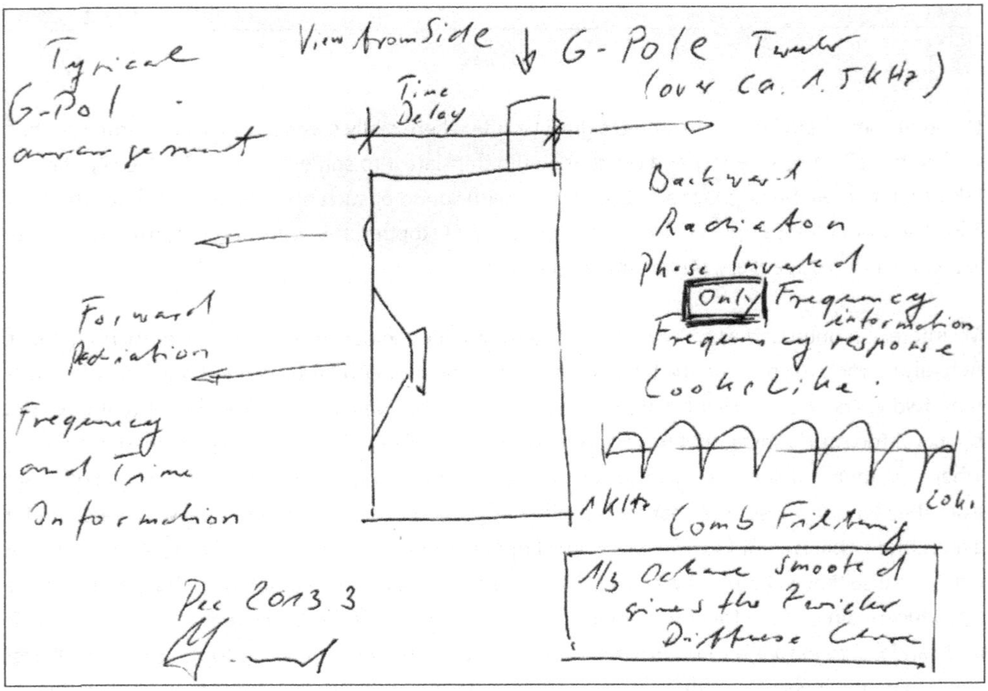

Figure 3

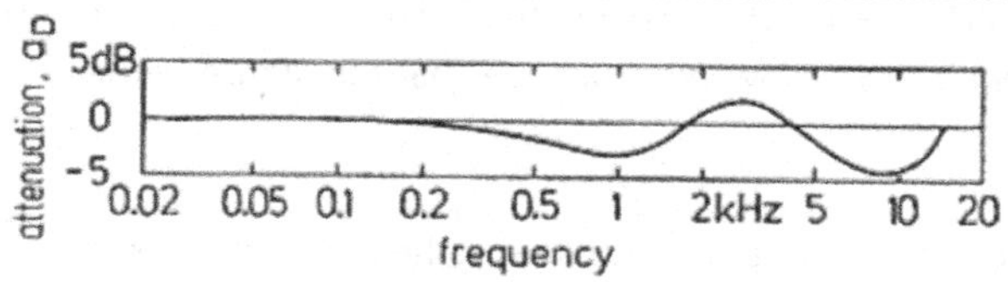

E. Zwicker, H. Fastl
Psychoacoustics -
Facts and Models, page 209
Springer 1999

Fig. 8.2. Attenuation, a_D, necessary to produce the same equal loudness of a pure tone in a diffuse and in a free sound field as a function of the pure tone's frequency

Figure 4

Orion sounds much better when he shapes the frontal response in a certain way so going away somewhat from a totally flat on axis response. Shortly after a tread appeared on diyaudio.com called "Flat is not correct". Looking at his website I found an interesting graph from Professor Zwicker (**figure 4**). It showed that the sensitivity of the human ear is different for direct and diffuse sound. I knew that to a certain degree at least since the late 80-ies. The Audio Physic Avanti that came out in 1989 had a set of resistors where you could tune the response to "Near field" or "Diffuse field". The diffuse field setting had more treble off axis for use in bigger, absorbing rooms. The idea came from using Stax headphones that also had a diffuse and free field setting but for other reasons. Must have being based on Zwickers research too but at that time I did not understand the full consequences. I thought long and hard about the Linkwitz solution to shape the response on axis. Actually I think he tries to shape the response because in a typical equi-distant triangle you hear the speaker not from straight ahead but some angle, say 35°, from the side.

Anyway, I concluded that this is not the solution I am searching for out of the simple notion that we need a norm that everybody agrees on. Imagine the control room of a typical recording studio. Usually the studio is well damped with a short and even reverberation time. The monitor speakers are usually setup to be flat at the control seat, sometimes with a "house curve" that tames the treble a bit. If the artist or the producer finds the sound unsatisfactory they will use equalization until the result is to their liking. Say, it is a Classical or Jazz production with unamplified natural instruments in a decent room they COULD put in the Linkwitz equalization, if they wish. Reproduced over the new Orion we would end up with "double equalization" if you understand what I mean.

Of course Siegfried is too wise to reshape the response in any drastic way and he has put in many hours of listening to arrive at a good compromise that works well on most of the material. Maybe I am simply "too German" to do it that way.

You will now ask when lightning struck me. Actually at exactly the point in time back then when I had finished the described thinking process. What would happen when I shaped the response flat on axis AND diffuse field equalized to the back of the dipole ? Bingo. Then lightning struck again. This time much harder because all kinds of memories came back from my earlier work and the meeting with Rick. What would happen if I destroy the impulse response to the back of my dipoles by using an additional tweeter out of phase with time delay and make sure that the diffuse frequency response is shaped after Zwicker?

I cannot describe my feeling but I felt that I had given birth to a baby: the G-Pole was born.

First I modified my big MPL baffles. I fitted a second ribbon horn in the treble and experimented with polarity, crossover and level. The Zwicker curve shows the biggest differences in the upper midrange and treble so I thought that adding the tweeter is enough. Also due to the Haas effect and steepening of the first transient in the ear, destroying the treble impulse response should be enough to avoid a second image from the back. Usually I am very measurement and simulation oriented but this time I just listened. Delaying the tweeter by some 20cm, inverting the polarity, giving it the same crossover as the front one and lowering the volume by a 3,9Ohm series resistor gave the desired result. Walking back and forth between the baffles, the sound from the back had approximately the same tonality but the second "phantom" image was gone.

I sat down and listened and was quite struck by the difference it made. Imaging and three-dimensionality was better and there was a sense of "air" I only know from the concert hall.
I was pretty satisfied. Since then I have demonstrated the effect to a lot to professionals and amateurs alike and 95% liked it a lot. Some of the population did not react to any significant degree but that where those kind of people like the mother of my son who love music but do not care much if it comes from a High End system or You Tube on a Mac Book. Lucky people. Saves lots of money and trouble.
The MPL is big and ugly (that must not be but I am simply too busy to build better looking baffles, the shoemaker always wears the worst shoes, you know) and I build it for myself. So I did not measure much but tuned it by ear to my taste and purpose.

When I design for other people I try to be more objective so I set up a plan how a G-Pole could be build that is smaller, cost efficient and well documented. It all ended up as the MiniMPL on diyaudio.com. I have shown some pictures and measurements there but did not publish the circuit diagram or any deeper instructions about its inner workings. This Guest Editorial gives me the opportunity and I am grateful for it.

Another chance came at the 2010 European Triode Festival in France. I had promised a lecture about the G-Pole principle. It was called "The G-Pole, a new radiation pattern".
When I got the program I was in shock! Nine AM on Saturday morning!
That is absolutely the wrong time for me. I work a lot at night because I do business with Japan and the USA and the time delay allows me to communicate with both countries at approximately the same time and Friday night is the only time in the week I take a rest, listen to my system or watch films until the morning. So that schedule hit me like a hammer.

Anyway, I thought let's give it a try. There will be only a few people anyway. Coming into the lecture room after at least *some* hours of sleep in the morning I got another shock. The room was packed! Fortunately I had put most of the slides on a USB stick and Martina Schoener and one of the French organizers (sorry I forgot the name, but you know who you are) did a wonderful job of helping me to find the correct pictures in the correct order. It was like telepathy but somehow I managed to struggle through the program. It was not my best lecture by a far margin and I witnessed a lot of skeptical faces so sorry boys and dolls. Next time I will make it more digestible.

What was positive was that a lot of people where now curious enough to listen to our system. Martina and I had set up the MiniMPL plus an active M-Dipole sub with a Garrard record player in a rather big room, even very close to the back wall where we put them on the first day after a long and dangerous car ride (it was quite foggy, windy and dark on the arrival day). We were simply too exhausted to tune the system in any way so this was a real acid test. Would the system sound good without the usual set up tricks and hours of tuning? It did, just fine. Once more the G-Pole principle had proven to be superbly tolerant to room acoustics and setup. A lot of people could not believe what that small tweeter hanging in the back on a spring could do and poor Holger Barske had to explain the principle to no one less than Frank Blöhbaum who asked his typical painful questions. Thanks Holger, you saved me once more. I was wasted and went to bed.

Very late on Saturday night I woke up and went back to the listening room. I had to throw some small stones against the windows to draw attention but was eventually let in. To my surprise there was a peaceful crowd around Martina and Holger listening for the 20st time to Paul Kuhn on Image records. The sound did not bother them at all after at least a 10 hour marathon and I thought "these G-Poles must be musical"....

An Essex-Echo Audio Tutorial: Analogue Equalization[1]

Malcolm Hawksford

0 Introduction

Audio systems commonly require frequency-response shaping networks. These can be used for response manipulation such as room correction, loudspeaker equalization and for systems employing complementary pre-emphasis and de-emphasis such as analogue tape systems and RIAA equalization used in phono preamplifiers. In addition frequency response shaping is widely employed for artistic and creative applications integrated either into the recording process or provided for end-user operation where response modification may be required.

In certain applications such as loudspeaker frequency response correction, the networks are fixed and specifically designed to match the inherent characteristics of the loudspeaker. The networks are synthesized to form an approximate complement to either or both the amplitude and the phase responses. Normally it is the amplitude response that takes precedence as phase response errors have lower subjective impact.

With general purpose equalization a flexible approach is required that enables either specific areas of the frequency spectrum to be targeted such as the bass region, the midrange or the treble region or much finer frequency discrimination in more critical applications. In the area of general-purpose equalization parametric or graphic equalizers are favoured while for broader band applications there are the classical 'tone controls' that focus on just the low-frequency and high-frequency regions to allow minor tonal corrections to be made.

Many equalizer topologies have their foundation forged in the analogue domain although these processes can easily be translated into digital realizations using the z-transform. Within the digital domain there is much greater freedom to control the interplay between amplitude response and phase response that can even be manipulated independently if required. It should be recalled that for the special case of minimum-phase processing, a change in amplitude response is always accompanied by a revised phase response as dictated by the Hilbert transform.

[1] This material was initiated during the 1980s and 1990s as examples of electronic equalization. The document is revised and offered as tutorial text which describes and analyses the underlying principles.

However, in this tutorial the focus is limited to three classical analogue equalization systems both to describe their underlying principles and also to show how the key parameters and components can be calculated. These are the series-feedback RIAA phono preamplifier with dc servo control, bass and treble 'tone controls' and the multi-band parametric equalizer. Each design is supported by a *Matlab* program (given in Appendices 1 to 3) which is suitable both as a design tool and to validate functionality. All designs assume idealized operational amplifiers or discrete equivalents where the circuits although presented in primitive form can readily be adapted into working prototypes.

1 Series-feedback RIAA phono preamplifier

Preamplifiers that amplify the signal derived either from a moving-coil (MC) or moving-magnet (MM) cartridge require embedded equalization defined by the RIAA. A Bode plot of this standard equalization characteristic together with its defining time constants is shown in **Figure 1-1**.

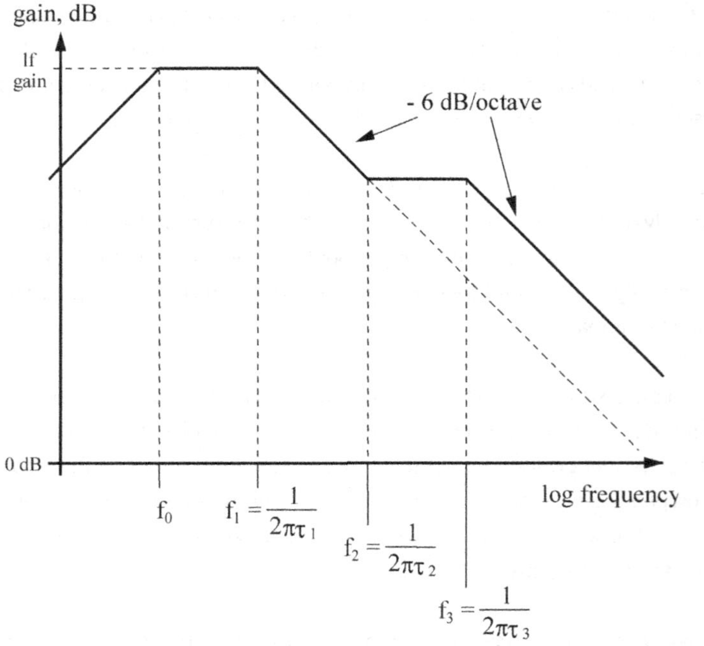

Figure 1-1 Bode plot of RIAA replay frequency response.

The RIAA equalization characteristic is actually a combination of two equalizers, one corrects for the magnetic cartridge and requires a simple response of -6 dB/octave to compensate for the cartridge having an output proportional to stylus velocity, while the second is a shelf filter which is designed to lower the amplitude of higher frequency signals during recording in order to improve tracking and reduce distortion, albeit with a signal-to-noise ratio penalty. There is also modest low-frequency pre-emphasis (applied in recording) and complementary de-emphasis (applied in playback) to allow the phono preamplifier to have a finite low-frequency gain and to reduce susceptibility to rumble and motor noise. The RIAA equalizer is defined mathematically by the transfer function $G_{RIAA}(f)$ and time constants, τ_1, τ_2 and τ_3 where,

$$G_{RIAA}(f) = G_{dc}\left\{\frac{(1 + j2\pi f\tau_2)}{(1 + j2\pi f\tau_1)(1 + j2\pi f\tau_3)}\right\}\left\{\frac{j2\pi f\tau_0}{(1 + j2\pi f\tau_0)}\right\} \quad (1\text{-}1)$$

In this equation G_{dc} is the arbitrary low-frequency gain, τ_1=3180μs, τ_2=318μs, τ_3=75μs, while τ_0 is an additional time constant defining the lower 3-dB break frequency f_0 which is typically set between 1 Hz and 20 Hz, where

$$\tau_0 = \frac{1}{2\pi f_0} \quad (1\text{-}2)$$

Figure 1-2 depicts the classic series-feedback phono-preamplifier topology which includes a non-inverting dc servo amplifier to set f_0 and which also forces a low output-offset voltage.

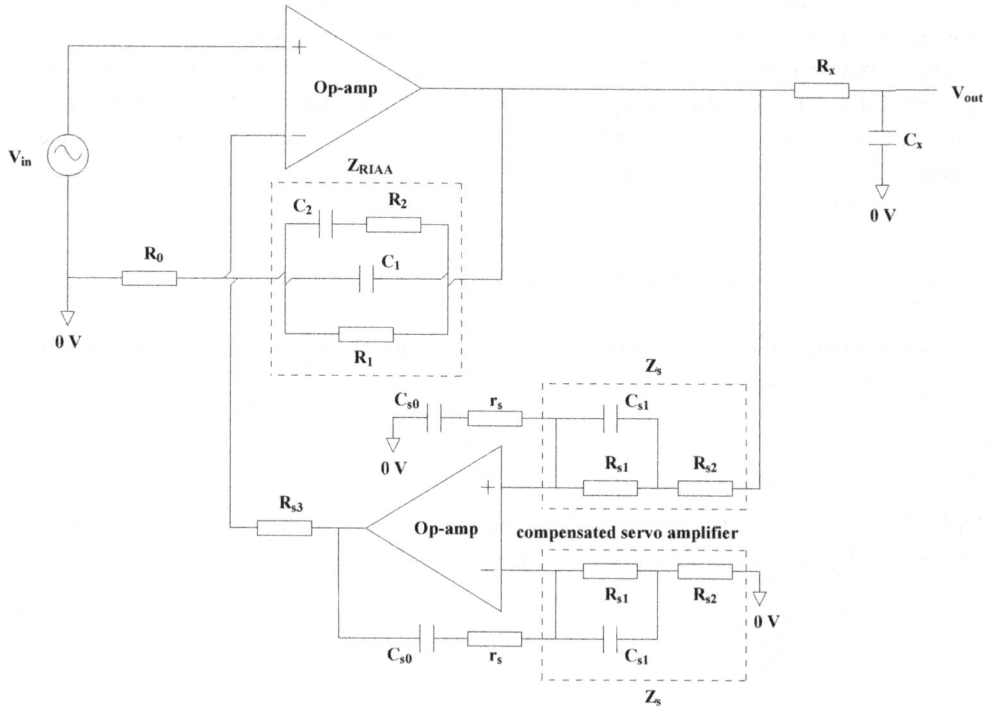

Figure 1-2 Series-feedback RIAA preamplifier with compensated servo amplifier.

The dc output-offset voltage of this amplifier is determined entirely by the input-offset voltage of the integrating servo amplifier where it is prudent to employ a Bi-FET or CMOS device to reduce input bias currents to extremely low levels, this way the voltage drop across each Z_S impedance is negligible allowing the differential input-offset voltage of the servo op-amp to dominate. If an extremely low output-offset voltage is required (say < 1 mV) then the servo op-amp should include its offset trim control, noting that the input-offset voltage of the main series-feedback amplifier is of no consequence in this configuration.

The principal design task for this amplifier is to match its frequency response as closely as possible to the RIAA target response defined by Equation 1-1. However, a minor complication and one not always recognized, is that when using the series-feedback configuration the first-order zero $(1 + j2\pi f\tau_2)$ within the RIAA network, Z_{RIAA} requires a minor modification to τ_2 which therefore will be designated τ_{m2}. Also the integrating servo amplifier demands frequency-response compensation to prevent error within the lower frequency range of the RIAA characteristic where this is enabled by impedance Z_S and resistance r_S.

Finally it should be recalled that at high frequency the gain-frequency response of the series-feedback amplifier (assuming an ideal op-amp) tends to unity whereas it should fall at -6 dB/octave in order to correctly match the RIAA response. To correct for this minor anomaly an additional first-order low-pass filter network is included by incorporating R_X and C_X. In Figure 1-2 this filter is shown positioned at the output of the preamplifier but it can be located anywhere in the amplifier chain providing its filtering function is retained, for example with an MC-preamplifier the cartridge source resistance R_S can be loaded by a capacitor C_L where $R_S C_L = R_X C_X = \tau_4$, this has the bonus of lowering ultrasonic and radio interference spuriae.

The design procedure is now developed for the circuit in Figure 1-2:

Let V_S be the output voltage of the non-inverting and integrating servo amplifier, the servo transfer function is then derived as,

$$\frac{V_s}{V_{out}} = \frac{1}{Z_s}\left(r_s + \frac{1}{j2\pi f C_{s0}}\right) = \frac{1}{Z_s}\left(\frac{1 + j2\pi f r_s C_{s0}}{j2\pi f C_{s0}}\right) \qquad (1\text{-}3)$$

Analyzing the main series-feedback amplifier by applying superposition (where $R_0//R_{S3}$ implies R_0 *in parallel with* R_{S3}), the output voltage V_{out} expressed in terms of V_{in} and V_S is,

$$V_{out} = \left\{V_{in}\left(1 + \frac{Z_{RIAA}}{R_0 // R_{s3}}\right) - V_s\frac{Z_{RIAA}}{R_s}\right\}\left\{\frac{1}{1 + j2\pi f\tau_4}\right\}$$

Substituting for V_S from Equation 1-3 and rearranging,

$$\frac{V_{out}}{V_{in}} = \left\{ \frac{1 + \dfrac{Z_{RIAA}}{R_0 // R_{s3}}}{1 + \dfrac{1}{Z_s}\left(\dfrac{1 + j2\pi f r_s C_{s0}}{j2\pi f C_{s0}}\right)\dfrac{Z_{RIAA}}{R_{s3}}\left(\dfrac{1}{1 + j2\pi f \tau_4}\right)} \right\}\left\{\frac{1}{1 + j2\pi f \tau_4}\right\} \qquad (1\text{-}4)$$

The RIAA defining impedance Z_{RIAA} requires the form,

$$Z_{RIAA} = \frac{R_1(1 + j2\pi f \tau_{m2})}{(1 + j2\pi f \tau_1)(1 + j2\pi f \tau_3)} \qquad (1\text{-}5)$$

Note that the dc resistance of this network is determined by resistor R_1 and that the time constant describing the zero $(1 + j2\pi f \tau_{m2})$ has been modified from τ_2 to τ_{m2}.

To facilitate pole-zero cancellation Z_S must take the form (see Figure 1-2),

$$Z_s = (R_{s1} + R_{s2})\frac{(1 + j2\pi f \tau_{m2})}{(1 + j2\pi f \tau_1)} \qquad (1\text{-}6)$$

and to cancel the pole associated with τ_3,

$$C_{s0} r_s = \tau_3 \qquad (1\text{-}7)$$

Hence substituting Z_{RIAA} from Equation 1-5 and Z_S from Equation 1-6,

$$\frac{Z_{RIAA}}{Z_s}\left\{\frac{1 + j2\pi f r_s C_{s0}}{j2\pi f C_{s0}}\right\}$$

$$\equiv \left\{\frac{R_1(1 + j2\pi f \tau_{m2})}{(1 + j2\pi f \tau_1)(1 + j2\pi f \tau_3)}\right\}\left\{\frac{1}{(R_{s1} + R_{s2})}\frac{(1 + j2\pi f \tau_1)}{(1 + j2\pi f \tau_{m2})}\right\}\left\{\frac{1 + j2\pi f r_s C_{s0}}{j2\pi f C_{s0}}\right\}$$

which by pole-zero cancellation reduces to,

$$\frac{Z_{RIAA}}{Z_s}\left\{\frac{1 + j2\pi f r_s C_{s0}}{j2\pi f C_{s0}}\right\} \equiv \frac{R_1}{(R_{s1} + R_{s2})}\left\{\frac{1}{j2\pi f C_{s0}}\right\}$$

Equation 1-4 can now be expressed,

$$\frac{V_{out}}{V_{in}} = \left\{ \frac{1 + \dfrac{Z_{RIAA}}{R_0 // R_{s3}}}{1 + \dfrac{R_1}{R_{s3}}\left\{\dfrac{1}{j2\pi f (R_{s1} + R_{s2}) C_{s0}}\right\}\left(\dfrac{1}{1 + j2\pi f \tau_4}\right)} \right\}\left\{\frac{1}{1 + j2\pi f \tau_4}\right\}$$

Observing $\tau_4 << C_{S0}(R_{S1} + R_{S2})$ the transfer function then reduces to its desired form,

$$\frac{V_{out}}{V_{in}} \approx \left\{1 + \frac{Z_{RIAA}}{R_0 // R_{s3}}\right\}\left\{\frac{j2\pi f\tau_0}{1 + j2\pi f\tau_0}\right\}\left\{\frac{1}{1 + j2\pi f\tau_4}\right\} \tag{1-8}$$

where, $\tau_0 = (R_{s1} + R_{s2})C_{s0}\dfrac{R_{s3}}{R_1}$ (1-9)

Hence the servo time constant to achieve a 3-dB break frequency f_0 is given by,

$$(R_{s1} + R_{s2})C_{s0} = \frac{R_1}{R_{s3}}\frac{1}{2\pi f_0} \tag{1-10}$$

Having now achieved an accurate estimate of the amplifier transfer function, Equation 1-8 can be equated to the target response specified by Equation 1-1,

$$G_{dc}\frac{(1 + j2\pi f\tau_2)}{(1 + j2\pi f\tau_1)(1 + j2\pi f\tau_3)}\left\{\frac{j2\pi f\tau_0}{(1 + j2\pi f\tau_0)}\right\} = \left\{1 + \frac{Z_{RIAA}}{R_0 // R_{s3}}\right\}\left\{\frac{j2\pi f\tau_0}{(1 + j2\pi f\tau_0)}\right\}\left\{\frac{1}{1 + j2\pi f\tau_4}\right\}$$

i.e. $$G_{dc}\frac{(1 + j2\pi f\tau_2)(1 + j2\pi f\tau_4)}{(1 + j2\pi f\tau_1)(1 + j2\pi f\tau_3)} = 1 + \frac{Z_{RIAA}}{R_0 // R_{s3}}$$

Substituting for Z_{RIAA} from Equation 1-5,

$$G_{dc}\frac{(1 + j2\pi f\tau_2)(1 + j2\pi f\tau_4)}{(1 + j2\pi f\tau_1)(1 + j2\pi f\tau_3)} \equiv 1 + \frac{R_1}{R_0 // R_{s3}}\frac{(1 + j2\pi f\tau_{m2})}{(1 + j2\pi f\tau_1)(1 + j2\pi f\tau_3)}$$

Rearranging,

$$G_{dc}\left(1 + j2\pi f(\tau_2 + \tau_4) - (2\pi f)^2\tau_2\tau_4\right) \equiv 1 + \frac{R_1}{R_0 // R_{s3}} + j2\pi f\left((\tau_1 + \tau_3) + \tau_{m2}\frac{R_1}{R_0 // R_{s3}}\right) - (2\pi f)^2\tau_1\tau_3 \tag{1-11}$$

Hence comparing terms it follows that,

$$G_{dc} = 1 + \frac{R_1}{R_0 // R_{s3}} \tag{1-12}$$

$$\tau_4 = \frac{\tau_1\tau_3}{G_{dc}\tau_2} \tag{1-13}$$

$$\tau_{m2} = \frac{G_{dc}}{(G_{dc} - 1)}\tau_2 + \frac{1}{(G_{dc} - 1)}\left(\frac{\tau_1\tau_3}{\tau_2} - \tau_1 - \tau_3\right) \tag{1-14}$$

Equation 1-12 is obvious from the circuit in Figure 1-2 while equations 1-13 and 1-14 allow τ_4 and τ_{m2} to be calculated once the dc-gain G_{dc} is set.

To complete the analysis the RIAA impedance Z_{RIAA} must be calculated to match the time constants τ_1, τ_{m2} and τ_3, also the compensation impedance Z_S in the servo must be determined according to Equation 1-6.

Analyzing Z_{RIAA} in Figure 1-2 gives,

$$Z_{RIAA} = \frac{R_1\left(1 + j2\pi f R_2 C_2\right)}{1 + j2\pi f\left(R_1C_2 + R_1C_1 + R_2C_2\right) - \left(2\pi f\right)^2 R_1C_1R_2C_2} \tag{1-15}$$

Substituting Z_{RIAA} from Equation 1-5,

$$\frac{R_1\left(1 + j2\pi f \tau_{m2}\right)}{1 + j2\pi f\left(\tau_1 + \tau_3\right) - \left(2\pi f\right)^2 \tau_1\tau_3} \equiv \frac{R_1\left(1 + j2\pi f R_2 C_2\right)}{1 + j2\pi f\left(R_1C_2 + R_1C_1 + R_2C_2\right) - \left(2\pi f\right)^2 R_1C_1R_2C_2}$$

Comparing terms,

$$R_2C_2 = \tau_{m2}$$

$$R_1C_2 + R_1C_1 + R_2C_2 = \tau_1 + \tau_3$$

$$R_1C_1R_2C_2 = \tau_1\tau_3$$

Hence assuming R_1 is given then C_1, C_2 and R_2 follows as,

$$C_1 = \frac{\tau_1\tau_3}{R_1\tau_{m2}} \tag{1-16}$$

$$C_2 = \frac{1}{R_1}\left(\tau_1 + \tau_3 - \tau_{m2} - \frac{\tau_1\tau_3}{\tau_{m2}}\right) \tag{1-17}$$

$$R_2 = R_1\frac{\tau_{m2}}{\left(\tau_1 + \tau_3 - \tau_{m2} - \frac{\tau_1\tau_3}{\tau_{m2}}\right)} \tag{1-18}$$

Analyzing the servo impedance Z_S in Figure 1-2,

$$Z_s = \left(R_{s1} + R_{s2}\right)\frac{\left(1 + j2\pi f \frac{R_{s1}R_{s2}}{\left(R_{s1} + R_{s2}\right)} C_{s1}\right)}{\left(1 + j2\pi f R_{s1} C_{s1}\right)} \tag{1-19}$$

Comparing terms in Equation 1-6 then,

$$R_{s1}C_{s1} = \tau_1 \quad \text{and} \quad \frac{R_{s1}R_{s2}}{(R_{s1} + R_{s2})}C_{s1} = \tau_{m2}$$

whereby,

$$C_{s1} = \frac{\tau_1}{R_{s1}} \tag{1-20}$$

$$R_{s2} = \frac{\tau_{m2}}{\tau_1 - \tau_{m2}}R_{s1} \tag{1-21}$$

From Equations 1-10, 1-7 and 1-21,

$$C_{s0} = \frac{R_1}{R_{s3}} \frac{1}{2\pi f_0 (R_{s1} + R_{s2})} \tag{1-22}$$

$$r_s = 2\pi f_0 \tau_3 (R_{s1} + R_{s2}) \frac{R_{s3}}{R_1} \tag{1-23}$$

while for the additional first-order high-frequency correction filter,

$$C_x = \frac{1}{R_x}\tau_4 \tag{1-24}$$

To calculate component values and validate the amplifier frequency response a *Matlab* program is presented in Appendix 1 incorporating a worked example where the input and derived values are:

Input data

$f_0 = 20\,Hz \quad R_1 = 110\,k\Omega \quad R_0 = 100\,\Omega \quad R_{s1} = 1\,M\Omega \quad R_{s3} = 10\,k\Omega$

$R_x = 1\,k\Omega$

Derived data

$R_2 = 15.915\,k\Omega \quad R_{s2} = 110.35\,k\Omega \quad C_1 = 6.8607\,nF \quad C_2 = 19.857\,nF$

$C_{s0} = 78.836\,nF \quad C_{s1} = 3.180\,nF \quad C_x = 674.46\,pF$

The program calculates the frequency response as shown in **Figure 1-3** both theoretically and through calculation using optimum component values given above, the responses both with and without the τ_4-filter correction are depicted.

To conclude the series-feedback phono preamplifier analysis the effect of frequency compensation in the integrating servo is explored in terms of error frequency responses as the question may be raised as to its significance. It should be observed as a validation of the derived equations for component values, that in terms of the design presented in Figure 1-2 and assuming optimum compo-

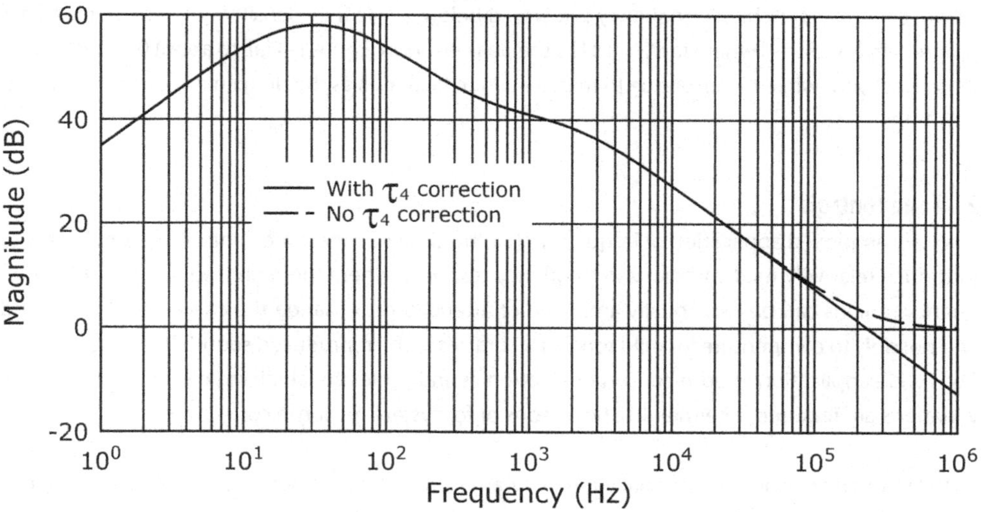

Figure 1-3 RIAA amplifier frequency response with and without τ_4-filter correction.

nent values and an ideal operational amplifier, the computed amplitude-response error was around 10^{-14} dB. In **Figure 1-4** the error response for $f_0 = 20$ Hz is shown when $r_s = 0$ as this is the least significant correction, as well as with $C_{s1} = 0$ and $r_s = 0$ so all compensation is removed and just a simple integrating servo remains. The third curve shows the error with the servo compensation included but with $\tau_{2m} = \tau_2$. Figure 1-4 shows r_s provides only minor correction and is probably of little consequence once component tolerances are taken into account, while the curve with $r_s = 0$ and $C_{s1} = 0$ reveals that compensation is definitely required. The last curve shows the effect of miscalculation.

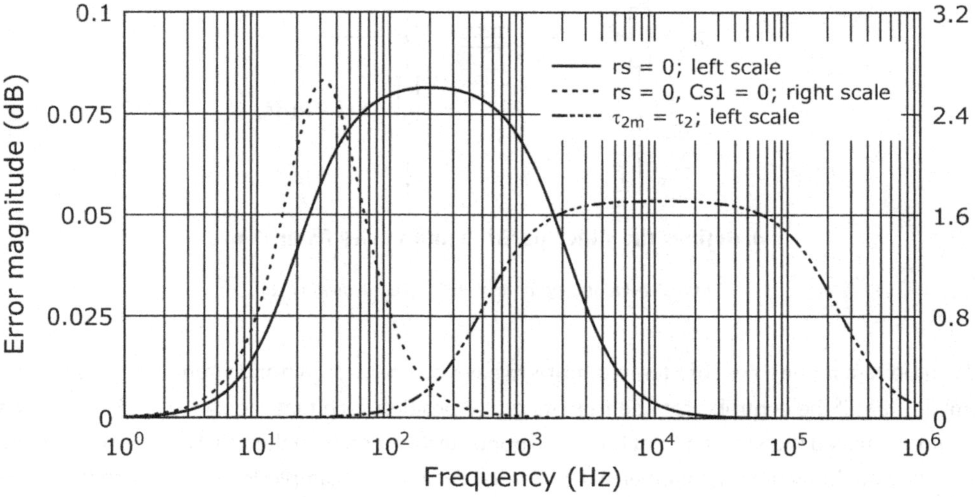

Figure 1-4 Error responses due to $r_s = 0$, $r_s = 0$ and $C_{s1} = 0$, and miscalculation with $\tau_{2m} = \tau_2$.

However, as the 3-dB break frequency is lowered below 20 Hz then sensitivity to compensation is progressively reduced where for $f_0 = 10$ Hz, the peak errors in Figure 1-4 fall to about 0.04 dB and 1.5 dB respectively, while the error due to miscalculation increases as the dc-gain is reduced, see Equation 1-14.

2 Tone controls

A widely employed application of frequency-shaping circuits is the 'tone control'. The aim here is to introduce relatively mild variations in frequency response where the high-frequency and low-frequency regions can be individually amplified or attenuated to change the character of the sound and possibly to compensate for mild tonal inaccuracies in the reproduced sound. This class of circuit finds wide application in both consumer electronics and professional mixing consoles where tonal variations are required for either artistic reasons or for system response correction.

As the frequency response is normally 'infinitely' adjustable, it is common to employ potentiometers (or their digital equivalent) to change the degree of high- and low-frequency boost or attenuation. As an aid to system analysis the notation illustrated in **Figure 2-1** is adopted. Here the potentiometer law is assumed linear and to have a total resistance R_X. The slider of the potentiometer then divides this resistance into two regions with respective resistances $0.5\,R_X\,(1 + m)$ and $0.5\,R_X\,(1 - m)$. Variable m represents the *control variable* (i.e. the position of the slider) and has a range -1 to 1 where $m = 0$ is the centre position. Also, note how the sum of the two resistor sections is obviously independent of m, that is $0.5\,R_X\,(1 + m) + 0.5\,R_X\,(1 - m) \equiv R_X$.

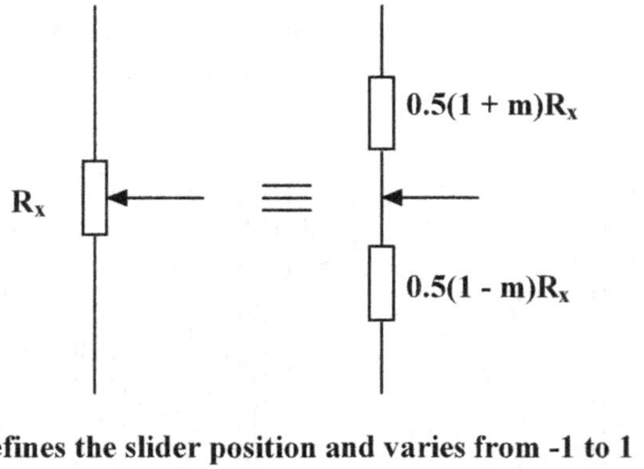

Figure 2-1 Variable potentiometer and equivalent circuit.

An important requirement of a tone control is to have symmetrical gain and attenuation control. This implies that if the control is rotated from its centre position such that the gain is X dB, then if the control is moved the same amount but in the opposite direction then the gain is –X dB. This may appear obvious but it does not happen necessarily by default. For example, let us assume that the gain of a circuit is proportional to potentiometer position so if $m = -0.5$, then the gain is 1.5 corresponding to 3.52 dB, while if $m = 0.5$ the gain is 0.5 corresponding to -6.02 dB, that is the gain law is asymmetric about $m = 0$.

The solution to this problem is shown conceptually in **Figure 2-2** where a unity-gain inverting amplifier has the potentiometer and a filter with transfer function λ positioned within a negative feedback path. Following the practice shown in Figure 2-1 the potentiometer is split into two sections linked by the control parameter m that represents the slider position. The system of Figure 2-2 is analysed as follows where in this example it is assumed that the filter network λ offers negligible loading on the potentiometer output seen on the slider.

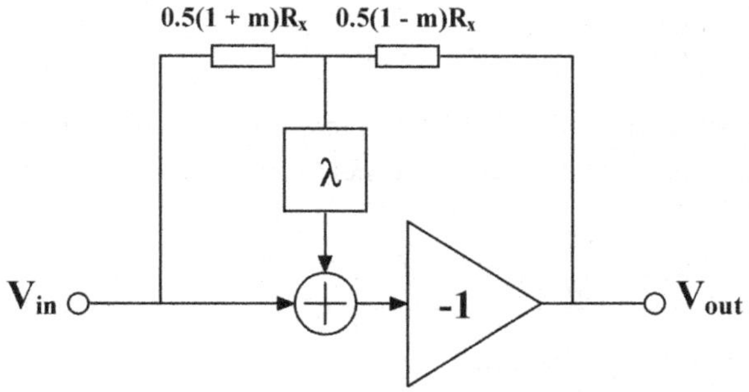

Figure 2-2 Concept of symmetrical gain control.

If the input voltage is V_{in} and the output voltage is V_{out}, then the output of λ is given as,

$$V_X = \lambda\left\{V_{in}\left(\frac{1-m}{2}\right) + V_{out}\left(\frac{1+m}{2}\right)\right\} \tag{2-1}$$

Considering signal flow within the system shown in Figure 2-2,

$$V_{out} = -V_X - V_{in}$$

Substituting for V_X,

$$V_{out} = -0.5\lambda\{V_{in}(1-m) + V_{out}(1+m)\} - V_{in}$$

Rearranging,

$$\frac{V_{out}}{V_{in}} = -\left[\frac{1 + 0.5\lambda(1-m)}{1 + 0.5\lambda(1+m)}\right] \tag{2-2}$$

Equation 2-2 reveals that gain law is symmetrical as a function of m and that for the special case of $m = 0$, the gain is precisely unity and independent of λ. This condition is mandatory for a tone control circuit because it is critical that there is a position of the potentiometer where the gain is independent of frequency, namely -1. Also, the condition of symmetry requires that for a given value of control parameter $m=M$, that

$$\left|\frac{V_{out}}{V_{in}}\right|_{m=M} = \left|\frac{V_{out}}{V_{in}}\right|_{m=-M}^{-1} \tag{2-3}$$

i.e. Equation 2-3 implies if $m = M$ the gain is X dB then for $m = -M$ the gain is –X dB.

In practice although the gain depends on m it is also frequency dependent where for a high-frequency tone control *or treble control,* the boost or cut in gain is applied only in the high-frequency range. This can be realised in the context of the system of Figure 2-2 by making λ a high-pass filter, for example taking the form

$$\lambda = \lambda_0\left(\frac{jf/f_0}{1 + jf/f_0}\right) \tag{2-4}$$

where λ_0 is the high-frequency gain and f_0 is the break frequency of the first-order high-pass filter. Substituting for λ in Equation 2-2 the overall transfer function is,

$$\left.\frac{V_{out}}{V_{in}}\right|_{hf} = -\left[\frac{1 + 0.5\lambda_0\left(\frac{jf/f_0}{1 + jf/f_0}\right)(1 - m)}{1 + 0.5\lambda_0\left(\frac{jf/f_0}{1 + jf/f_0}\right)(1 + m)}\right]$$

that is,

$$\left.\frac{V_{out}}{V_{in}}\right|_{hf} = -\left[\frac{1 + \frac{jf}{f_0}(1 + 0.5\lambda_0(1 - m))}{1 + \frac{jf}{f_0}(1 + 0.5\lambda_0(1 + m))}\right] \tag{2-5}$$

from which the extreme low-frequency and high-frequency gains are respectively,

$$\left.\frac{V_{out}}{V_{in}}\right|_{f\to 0} = -1 \quad \text{and} \quad \left.\frac{V_{out}}{V_{in}}\right|_{f\to\infty} = -\left[\frac{1 + 0.5\lambda_0(1 - m)}{1 + 0.5\lambda_0(1 + m)}\right] \tag{2-6(a,b)}$$

A similar approach can be taken to the *bass control* design that can be implemented by making a low-pass filter defined as,

$$\lambda = \lambda_0\left(\frac{1}{1 + jf/f_0}\right) \tag{2-7}$$

This results in the transfer function,

$$\left.\frac{V_{out}}{V_{in}}\right|_{lf} = -\left[\frac{1 + 0.5\lambda_0(1 - m)}{1 + 0.5\lambda_0(1 + m)}\right]\left[\frac{1 + j\frac{f/f_0}{1 + 0.5\lambda_0(1 - m)}}{1 + j\frac{f/f_0}{1 + 0.5\lambda_0(1 + m)}}\right] \tag{2-8}$$

Here the extreme low-frequency and high-frequency gains are reversed implying functional symmetry in the frequency domain,

$$\left.\frac{V_{out}}{V_{in}}\right|_{f\to 0} = -\left[\frac{1 + 0.5\lambda_0(1 - m)}{1 + 0.5\lambda_0(1 + m)}\right] \quad \text{and} \quad \left.\frac{V_{out}}{V_{in}}\right|_{f\to\infty} = -1 \qquad \text{(2-9(a,b))}$$

The system described in Figure 2-2 may be extended to include a larger number (say *N*) of control functions as shown in **Figure 2-3**.

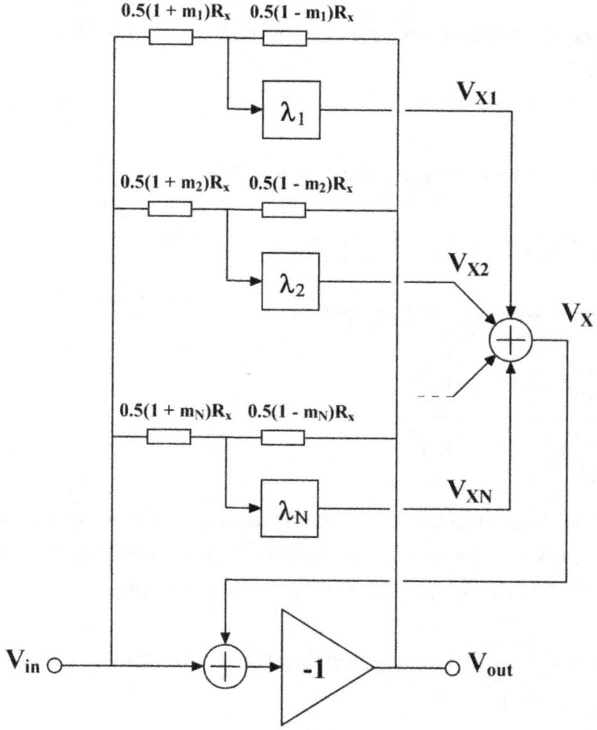

Figure 2-3 Multi-band control topology.

The transfer function of the multi-band filter of Figure 2-3 is analysed as follows:

Let,

$$V_{X1} = 0.5\lambda_1\{V_{in}(1 - m_1) + V_{out}(1 + m_1)\}$$

$$V_{X2} = 0.5\lambda_2\{V_{in}(1 - m_2) + V_{out}(1 + m_2)\}$$

up to

$$V_{XN} = 0.5\lambda_N\{V_{in}(1 - m_N) + V_{out}(1 + m_N)\}$$

The composite filtered signal is formed by summation over N as,

$$V_X = \sum_{r=1}^{N}\left[0.5\lambda_r\left\{V_{in}\left(1 - m_r\right) + V_{out}\left(1 + m_r\right)\right\}\right] \tag{2-10}$$

that is,

$$V_X = V_{in}\sum_{r=1}^{N}0.5\lambda_r\left(1 - m_r\right) + V_{out}\sum_{r=1}^{N}0.5\lambda_r\left(1 + m_r\right)$$

Hence, by considering signal flow within the system shown in Figure 2-3,

$$V_{out} = -V_X - V_{in}$$

and substituting for V_X then the output voltage V_{out} follows as,

$$V_{out} = -V_{in}\sum_{r=1}^{N}\left[0.5\lambda_r\left(1 - m_r\right)\right] - V_{out}\sum_{r=1}^{N}\left[0.5\lambda_r\left(1 + m_r\right)\right] - V_{in}$$

Rearranging to determine the overall transfer function,

$$\frac{V_{out}}{V_{in}} = -\frac{1 + \sum_{r=1}^{N}\left[0.5\lambda_r\left(1 - m_r\right)\right]}{1 + \sum_{r=1}^{N}\left[0.5\lambda_r\left(1 + m_r\right)\right]} \tag{2-11}$$

A practical two-band tone control implementation based upon the general structure of Figure 2-3 is shown in **Figure 2-4** for $N = 2$ where both high-frequency and low-frequency first-order filter sections have been included to form a combined bass and treble tone control.

To analyse this circuit the overall transfer function is derived in general terms by putting $N = 2$ in equation 2-8, giving

$$\frac{V_{out}}{V_{in}} = -\frac{1 + 0.5\lambda_1\left(1 - m_1\right) + 0.5\lambda_2\left(1 - m_2\right)}{1 + 0.5\lambda_1\left(1 + m_1\right) + 0.5\lambda_2\left(1 + m_2\right)} \tag{2-12}$$

where m_1, λ_1 are assigned to the bass control and m_2, λ_2, to the treble control. The next stage is to determine the filter transfer functions λ_1 and λ_2.

Bass control transfer function λ_1

The bass control circuit requires λ_1 being a low-pass filter as the influence of this control function must diminish with increasing frequency. The main amplifier used in Figure 2-4 is a virtual earth, shunt-feedback stage; therefore the impedance observed between the inverting input and ground is low tending to zero in the idealized case. In Figure 2-3 a summing amplifier was required whilst in the practical circuit this is achieved by resistive mixing and a virtual earth amplifier. The output of fil-

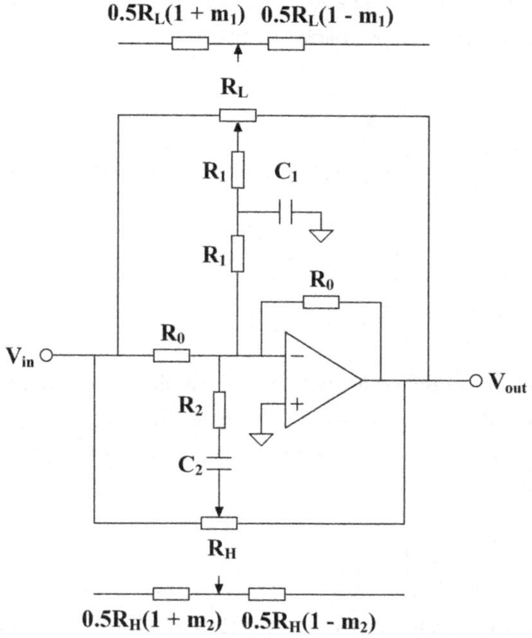

Figure 2-4 High-frequency and low-frequency 'tone controls'.

ter λ_1 is effectively at the node formed by the two resistors R_1 and capacitor C_1 and the summing stage when viewed in the context of the shunt-feedback amplifier has gain R_0/R_1. A model of filter λ_1 and the effective gain is shown in **Figure 2-5**.

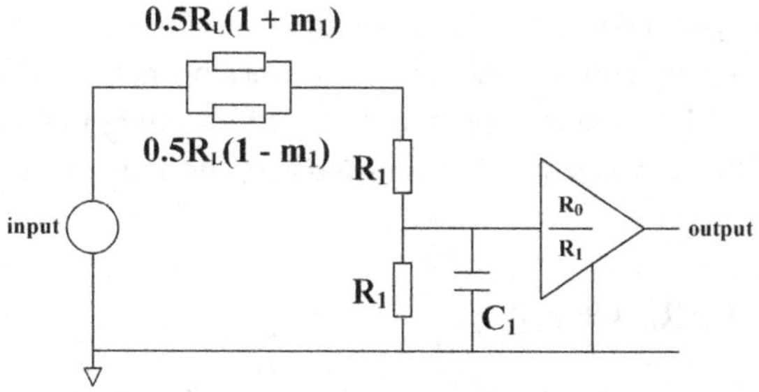

Figure 2-5 Model of bass control filter λ_1.

The input voltage to the filter is derived from a mix of V_{in} and V_{out} as shown in Figure 2-3. However, in analyzing the filter circuit the Thévenin source resistance of the potentiometer must be included which is equal to the parallel combination of the two resistive sections of the potentiometer, assuming the shunt-feedback amplifier has negligible output impedance. This is shown in Figure 2-5 by including resistance $0.5\,R_L\,(1 + m_1)\,//\,0.5\,R_L\,(1 - m_1)$ in series with R_1, where R_L is the bass-potentiometer resistance.

Having reconfigured the filter circuit and assuming an idealised operational amplifier in the shunt-feedback stage, transfer function λ_1 follows by the potential divider rule as,

$$\lambda_1 = \frac{R_0}{R_1}\left(\frac{\dfrac{R_1}{1 + j\omega R_1 C_1}}{\dfrac{R_1}{1 + j\omega R_1 C_1} + 0.25R_L(1 + m_1)(1 - m_1) + R_1}\right)$$

which simplifies to,

$$\lambda_1 = \frac{R_0}{2R_1}\left\{1 + 0.125\frac{R_L}{R_1}(1 + m_1)(1 - m_1) + \left(0.125\frac{R_L}{R_1}(1 + m_1)(1 - m_1) + 0.5\right)j\omega R_1 C_1\right\}^{-1} \quad (2\text{-}13)$$

This expression may be simplified further if the Thévenin equivalent resistance of the potentiometer is a small fraction of the resistance R_1. This implies that as the bass control is changed, the 3-dB break frequency of the low-pass filter λ_1 does not change. In this case the transfer function reduces to its approximate form,

$$\lambda_1 = \frac{R_0}{2R_1}\left(\frac{1}{1 + 0.5j\omega R_1 C_1}\right) \quad (2\text{-}14)$$

Equation 2-14 confirms the filter has a low-pass response with a dc-gain 0.5 R_0 / R_1 and with a time constant 0.5 R_1C_1 formed by $R_1//R_1//C_1$.

Treble-control transfer function λ_2

For the treble-control circuit λ_2 is a high-pass filter as the effect of this control function must diminish with decreasing frequency. A similar analysis procedure to the bass control is followed but now the filter is a high-pass filter and formed by R_2 and C_2. The junction of R_2 and C_2 can be considered as the output of the high-pass filter since this node forms one of the inputs to the shunt-feedback summing amplifier, hence assuming an idealised operational amplifier, the gain is R_0 / R_2. **Figure 2-6** shows an analytical model for determining filter transfer function λ_2 that includes the Thévenin equivalent resistance of the treble control R_H expressed in terms of the boost and cut control variable m_2

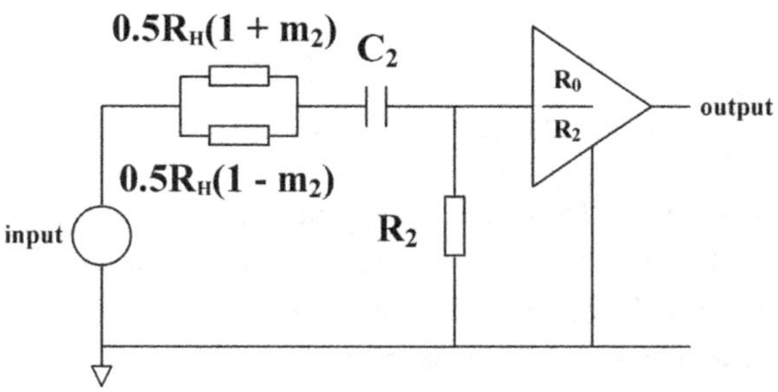

Figure 2-6 Model of treble control filter λ_2.

Considering the first-order high-pass filter shown in Figure 2-6 as a 'potential divider' network cascaded with a gain stage R_0 / R_2, then λ_2 follows as,

$$\lambda_2 = \frac{R_0}{R_2}\left(\frac{R_2}{R_2 + 0.25R_H(1+m_2)(1-m_2) + \dfrac{1}{j\omega C_2}}\right)$$

which simplifies to,

$$\lambda_2 = \frac{R_0}{R_2}\left(\frac{j\omega R_2 C_2}{1 + j\omega\{R_2 + 0.25R_H(1+m_2)(1-m_2)\}C_2}\right) \tag{2-15}$$

For the case where the Thévenin output resistance of potentiometer R_H is much less than resistor R_2, then Equation 2-15 reduces to its approximate form,

$$\lambda_2 = \frac{R_0}{R_2}\left(\frac{j\omega R_2 C_2}{1 + j\omega R_2 C_2}\right) = \frac{j\omega R_0 C_2}{1 + j\omega R_2 C_2} \tag{2-16}$$

Having now determined the individual tone-control filter transfer functions λ_1 and λ_2, the approximate (i.e. neglecting potentiometer resistances and assuming an ideal op-amp) overall transfer function can be derived from Equation 2-12 by substituting for λ_1 and λ_2 from Equation 2-14 and 2-16, whereby

$$\frac{V_{out}}{V_{in}} = \frac{1 + \left(\dfrac{1-m_1}{2}\right)\left\{\dfrac{R_0}{2R_1}\dfrac{1}{(1+0.5j\omega R_1 C_1)}\right\} + \left(\dfrac{1-m_2}{2}\right)\left\{\dfrac{j\omega R_0 C_2}{1+j\omega R_2 C_2}\right\}}{1 + \left(\dfrac{1+m_1}{2}\right)\left\{\dfrac{R_0}{2R_1}\dfrac{1}{(1+0.5j\omega R_1 C_1)}\right\} + \left(\dfrac{1+m_2}{2}\right)\left\{\dfrac{j\omega R_0 C_2}{1+j\omega R_2 C_2}\right\}} \tag{2-17}$$

To plot an example range of frequency responses the *Matlab* routine presented in Appendix 2 was used where the gain-attenuation of the low-frequency and high-frequency controls both span -5 dB to 5 dB. **Figure 2-7** presents a set of frequency responses with identical high- and low-frequency gain ranges by making,

$$R_2 = 2R_1 \tag{2-18}$$

and,

$$R_0 = 2R_1\left\{10^{\left(\frac{Maximum\ gain|_{dB}}{20}\right)} - 1\right\}_{m_1, m_2 = -1} \tag{2-19}$$

m_1 and m_2 range from -1 to 1 in 0.2 increments. The results confirm a symmetrical boost and cut characteristic on a dB-logarithmic frequency scale together with well defined corner frequencies with gain increments that are approximately equi-interval on a dB scale.

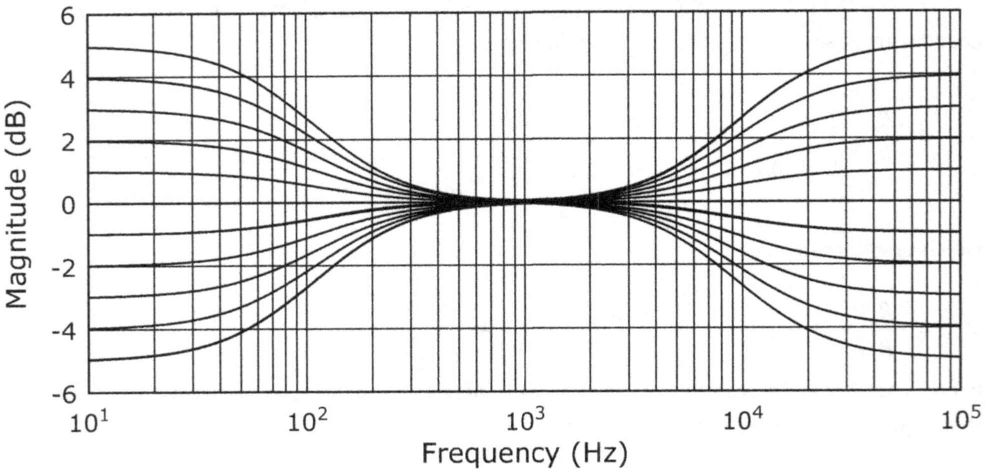

Figure 2-7 Example frequency response plots for the circuit shown in Figure 2-4.

3 Parametric equalizers

A parametric equalizer is similar conceptually to the 'tone-control' circuit described in *Section 2* although the bandwidth is usually more constrained being characterized by a band-pass response rather than a low-pass or high-pass response. In this scheme there are normally multiple bands *N* (e.g. $N = 5$ to 10) where gain, frequency and bandwidth of each filter are specified and as such form the control *parameters* of a parametric equalizer. **Figure 3-1** shows a structure using passive band-pass filters that is based upon the topology shown in **Figure 2-3** and can be analysed using Equation 2-11 together with appropriate filter definitions.

The filters for each band-pass section can be determined with reference to the circuit shown in Figure 3-2 that depicts one λ_r-filter where again the operational amplifier is assumed to be ideal and the gain control *VR*, is represented by its Thévenin equivalent circuit. Following a similar procedure to Section 2, analysis of the λ_r-filter gives,

$$\lambda_r = \frac{R_0}{R_r}\left\{\frac{j\omega R_r C_r}{1 + j\omega C_r\left(R_r + 0.25VR_r\left(1 + m_r\right)\left(1 - m_r\right)\right) + \left(j\omega\right)^2 L_r C_r}\right\} \tag{3-1}$$

If the Thévenin resistance of the potentiometer VR_r is negligible then Equation 3-1 reduces to,

$$\lambda_r = \frac{j\omega R_0 C_r}{1 + j\omega R_r C_r + \left(j\omega\right)^2 L_r C_r} \tag{3-2}$$

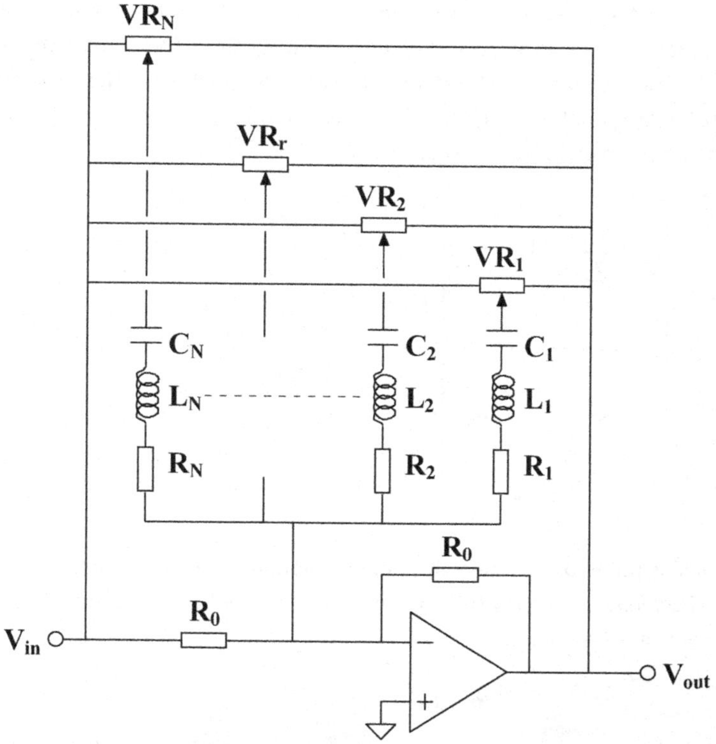

Figure 3-1 N-band parametric equalizer with passive band-pass filter sections.

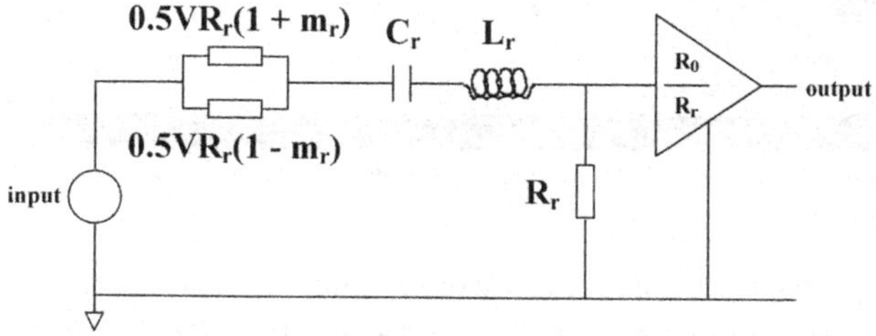

Figure 3-2 Band-pass λ_r-filter used in parametric equalizer.

Equations 3-1 and 3-2 represent the classic second-order band-pass filter that can be expressed generically in terms of mid-band gain G_r, quality factor Q_r and un-damped natural resonance ω_r,

$$\lambda_r = G_r\left(\frac{j\dfrac{1}{Q_r}\dfrac{\omega}{\omega_r}}{1 + j\dfrac{1}{Q_r}\dfrac{\omega}{\omega_r} + \left(\dfrac{j\omega}{\omega_r}\right)^2}\right) \tag{3-3}$$

As such parameters G_r, Q_r and ω_r relate to the user-controls of the parametric equalizer, these controls being repeated for each channel of the *N*-band equalizer. They control the cut and boost, bandwidth and centre frequency respectively for each individual band. Hence, substituting for λ_r from Equation 3-3 into Equation 2-11, then the general transfer function for an *N*-band parametric equalizer of the form shown in Figure 3-1 is given as,

$$\frac{V_{out}}{V_{in}} = -\left\{ \frac{1 + \sum_{r=1}^{N} G_r \left(\frac{1 - m_r}{2} \right) \left(\dfrac{j \dfrac{1}{Q_r} \dfrac{\omega}{\omega_r}}{1 + j \dfrac{1}{Q_r} \dfrac{\omega}{\omega_r} + \left(\dfrac{j\omega}{\omega_r} \right)^2} \right)}{1 + \sum_{r=1}^{N} G_r \left(\frac{1 + m_r}{2} \right) \left(\dfrac{j \dfrac{1}{Q_r} \dfrac{\omega}{\omega_r}}{1 + j \dfrac{1}{Q_r} \dfrac{\omega}{\omega_r} + \left(\dfrac{j\omega}{\omega_r} \right)^2} \right)} \right\} \qquad (3\text{-}4)$$

By way of illustration of a commercial 4-band parametric equalizer control surface is shown in **Figure 3-3** (*G-Sonique equalizer*) where cut and boost control, bandwidth (i.e. Q-control) and centre frequency control are shown.

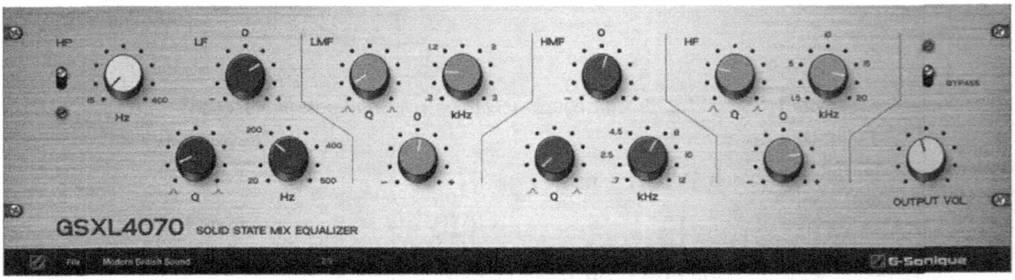

Figure 3-3 Front panel of commercial parametric equalizer (G-Sonique equalizer).

As an example of the parametric filter capabilities a family of frequency responses is plotted in **Figure 3-4** for a 5-band system where centre frequencies are 50 Hz, 250 Hz, 1 kHz, 5 kHz and 15 kHz, respective Q-factors are 10, 3, 5, 1, 10 and $m_r|_{r=1to5}$ = -1 to 1 in 0.1 increments for a gain range -15 dB to 15 dB for each band *r*. The *Matlab* routine used to model this parametric equalizer is presented in Appendix 3.

In a practical parametric equalizer it is normal to use active filters to synthesise each band-pass filter section where commonly a bi-quadratic structure enables the desired parametric control flexibility. This structure has the advantage that it is straightforward to vary the centre frequency and Q-factor with orthogonal control. Orthogonal control implies that when for example the centre frequency is changed then the Q-factor remains constant and vice versa, where a basic feedback system is shown in **Figure 3-5**.

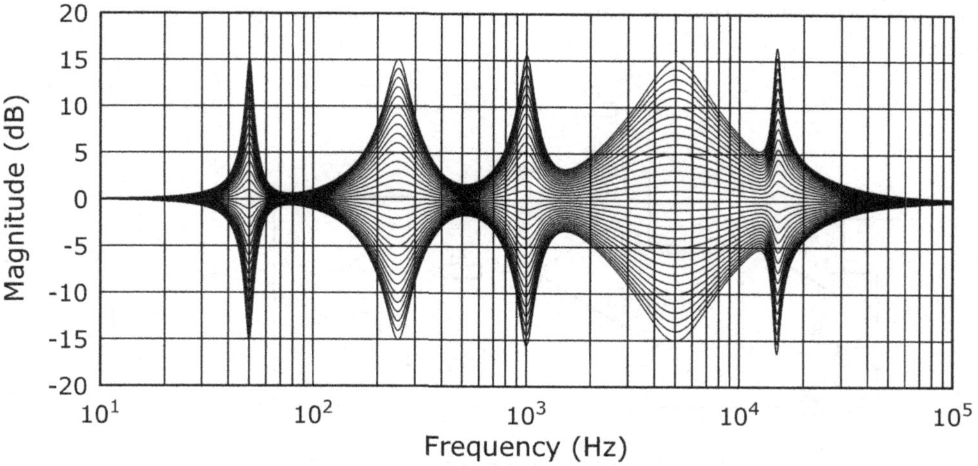

Figure 3-4 Example frequency response plots for a 5-band parametric equalizer.

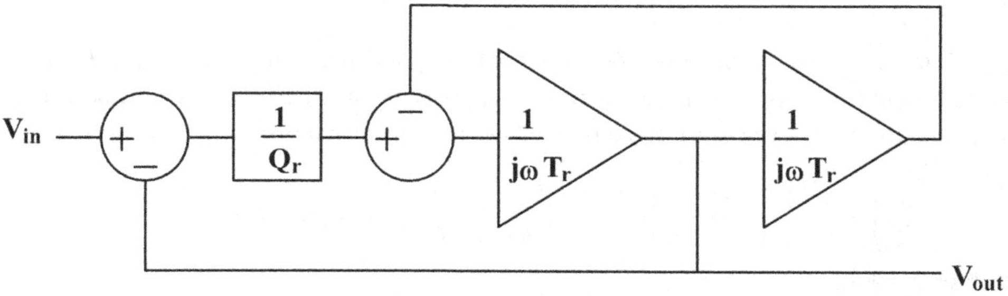

Figure 3-5 Band-pass filter feedback system with parametric control.

The transfer function of the filter shown in Figure 3-5 is given by,

$$\left.\frac{V_{out}}{V_{in}}\right|_r = \frac{\frac{1}{Q_r} j\omega T_r}{1 + \frac{1}{Q_r} j\omega T_r + \left(j\omega T_r\right)^2} \equiv \frac{\frac{1}{Q}\left(\frac{j\omega}{\omega_r}\right)}{1 + \frac{1}{Q_r}\left(\frac{j\omega}{\omega_r}\right) + \left(\frac{j\omega}{\omega_r}\right)^2} \tag{3-5}$$

where $\omega_r = 1/T_r$. In this system the time constants T_r of the two integrators must change in synchronism to set the undamped natural resonant frequency while changing the gain of the box labelled $1/Q_r$ directly controls the Q-factor. It is evident from Equation 3-5 that whatever Q-factor is selected the maximum gain between input and output is unity. This strategy lowers the probability of overload distortion either at the output or internally within the circuit where the filter shown in Figure 3-5 is configured to circumvent this effect.

In **Figure 3-6** the filter of Figure 3-5 has been translated into a practical operational amplifier-based circuit.

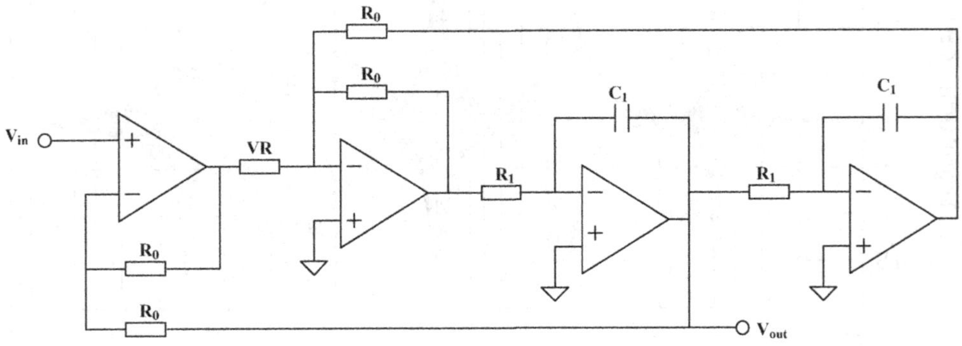

Figure 3-6 Bi-quadratic band-pass filter with parametric control.

The bi-quadratic filter can be analysed as follows:

If V_{out} is the band-pass output, then the output of the second inverting integrator is $-j\omega R_1 C_1 V_{out}$ while the input to the first inverting integrator is $-V_{out}/j\omega R_1 C_1$. If {VR} is a variable resistor that sets the Q-factor, then observing the first-stage amplifier and Q-factor control stage, it follows that,

$$\left\{V_{in}\left(1 + \frac{R_0}{R_0}\right) - V_{out}\left(\frac{R_0}{R_0}\right)\right\}\left(-\frac{R_0}{\{VR\}}\right) - \frac{V_{out}}{j\omega R_1 C_1}\left(-\frac{R_0}{R_0}\right) = V_{out}\left(-j\omega R_1 C_1\right)$$

which simplifies to,

$$\frac{V_{out}}{V_{in}} = \frac{2\left(\frac{R_0}{\{VR\}}\right) j\omega R_1 C_1}{1 + \left(\frac{R_0}{\{VR\}}\right) j\omega R_1 C_1 + \left(j\omega R_1 C_1\right)^2} \equiv \frac{2\frac{1}{Q_r}\left(\frac{j\omega}{\omega_r}\right)}{1 + \frac{1}{Q_r}\left(\frac{j\omega}{\omega_r}\right) + \left(\frac{j\omega}{\omega_r}\right)^2} \tag{3.6}$$

where $\omega_r = 1/(R_1 C_1)$ and $Q_r = \{VR\} / R_0$ noting Q_r is directly proportional to {VR}.

Individual bi-quadratic filter sections are used for each parametric stage and replace the set of RLC series filters shown in Figure 3-1. Also practical circuits are often DC-coupled so a final design may incorporate a servo amplifier to steer the output voltage to zero. Other variations include options for balanced input and outputs where these find favour in professional applications to minimise hum and interference induction that can occur in long cable runs typically found in broadcast and recording studios.

Editor's note

The following appendices will be posted on the Linear Audio website under 'Articles'.

Appendix 1 Matlab routine for series-feedback phono preamplifier

```
% RIAA_series_feedback.m
% 04.08.13
% series feedback RIAA preamp with servo
% preferred values can be entered in lines 49 to 60
home; clc; clear; colordef white; close
fprintf('SERIES FEEDBACK RIAA PREAMP WITH SERVO\n')

% set ww = 1 to write Excel file of frequency responses
ww=0;

% RIAA data
T1=3180*10^-6; % RIAA EQ time constant, second
T2=318*10^-6; % RIAA EQ time constant, second
T3=75*10^-6; % RIAA EQ time constant, second
f0=20; % 3-dB break frequency, Hz

% frequency vector (1 Hz to 100 kHz, N points/dec)
N=1000;
f=10.^(0:1/N:6);

% component values
R1=110000; % RIAA dc resistor, ohm
R0=100; % feedback resistor to ground, ohm
Rs1=10^6; % servo resistor, ohm
Rs3=10000; % series output resistor of servo
Rx=1000; % resistance of RC filter associate with T4

% response correction for series feedback topology
Gdc=1+R1/(R0*Rs3/(R0+Rs3)); % dc gain neglecting servo
Tm2=(Gdc*T2+T1*T3/T2-T1-T3)/(Gdc-1);
T4=T1*T3/(Gdc*T2);

% series feedback RIAA preamp
C1=(T1*T3/Tm2)/R1;
C2=(T1+T3-Tm2-T1*T3/Tm2)/R1;
R2=R1*Tm2/(T1+T3-Tm2-T1*T3/Tm2);
```

```
% servo amplifier components
T0=1/(2*pi*f0);
Rs2=Rs1*Tm2/(T1-Tm2);
Cs0=(R1/Rs3)/(2*pi*f0*(Rs1+Rs2));
Cs1=T1/Rs1;
rs=2*pi*f0*T3*(Rs1+Rs2)*Rs3/R1;

% low-pass correction filter, T4
Cx=T4/Rx;

% enter preferred values
R0=R0; %#ok<ASGSL>
R1=R1; %#ok<ASGSL>
R2=R2; %#ok<ASGSL>
C1=C1; %#ok<ASGSL>
C2=C2; %#ok<ASGSL>
Rs1=Rs1; %#ok<ASGSL>
Rs2=Rs2; %#ok<ASGSL>
Rs3=Rs3; %#ok<ASGSL>
Cs0=Cs0; %#ok<ASGSL>
Cs1=Cs1; %#ok<ASGSL>
Rx=Rx; %#ok<ASGSL>
Cx=Cx; %#ok<ASGSL>

% computed (or preferred value) data for series feedback preamplifier
fprintf('\n\nRIAA EQ COMPONENTS FOR SERIES FEEDBACK MC PRE-AMPLIFIER\n')
R0 %#ok<NOPTS>
R1 %#ok<NOPTS>
R2 %#ok<NOPTS>
C1 %#ok<NOPTS>
C2 %#ok<NOPTS>
Rs1 %#ok<NOPTS>
Rs2 %#ok<NOPTS>
Rs3 %#ok<NOPTS>
Cs0 %#ok<NOPTS>
Cs1 %#ok<NOPTS>
Rx %#ok<NOPTS>
Cx %#ok<NOPTS>
```

```
% plot reference RIAA frequency response
Griaa=abs(Gdc*(1+i*2*pi*f*T2)./((1+i*2*pi*f*T1).*(1+i*2*pi*f*T3)).*(i*2*pi*f*T0./(1+i*2*pi*f*T0)));
semilogx(f,20*log10(Griaa),'k')
title('Reference RIAA frequency response')
xlabel('Frequency, Hz')
ylabel('Amplitude, dB')
grid; pause(3); close

% theoretical frequency response for RIAA TF with and without T0 correction
G0=abs((1+(Gdc-1)*(1+i*2*pi*f*Tm2).*
(i*2*pi*f*T0./(1+i*2*pi*f*T0))./((1+i*2*pi*f*T1).*(1+i*2*pi*f*T3)))./(1+i*2*pi*f*T4));
G1=abs(1+(Gdc-1)*(1+i*2*pi*f*Tm2).*
(i*2*pi*f*T0./(1+i*2*pi*f*T0))./((1+i*2*pi*f*T1).*(1+i*2*pi*f*T3)));
semilogx(f,20*log10(G0),'b')
hold
semilogx(f,20*log10(G1),'r')
title('Theoretical series-feedback RIAA frequency response with (blue) and without (red) T4')
xlabel('Frequency, Hz')
ylabel('Amplitude, dB')
grid; pause(3); close

% frequency response calculations using component values
Zriaa=R1*(1+i*2*pi*f*R2*C2)./(1+i*2*pi*f*(R1*C2+R1*C1+R2*C2)-(2*pi*f).^2*R1*C1*R2*C2);
Zs=(Rs1+Rs2).*(1+i*2*pi*f*Rs1*Rs2/(Rs1+Rs2)*Cs1)./(1+i*2*pi*f*Rs1*Cs1);
servo=(rs+(i*2*pi*f*Cs0).^-1)./Zs;
Gamp=(1+Zriaa/(R0*Rs3/(R0+Rs3)))./(1+i*2*pi*f*Rx*Cx);
Gamp=Gamp./(1+servo.*Zriaa/Rs3);

% frequency response derived from series feedback amp
semilogx(f,20*log10(abs(Gamp)),'k')
title('Series feedback amplifier RIAA frequency response')
xlabel('Frequency, Hz')
ylabel('Amplitude, dB')
grid; pause(3); close

% error frequency response derived from series feedback amp
semilogx(f,20*log10(abs(Gamp./Griaa)),'k')
title('Series feedback amplifier RIAA error frequency response')
xlabel('Frequency, Hz')
ylabel('Amplitude, dB')
grid; pause(.5);
```

```
% write Excel file for frequency responses
if ww==1 % write Excel file
xlswrite('RIAA_frequency_response', (f;20*log10(G0);20*log10(G1))');
xlswrite('error_frequency_response', (f;20*log10(abs(Gamp./Griaa)))');
end
pause; close; % close graphs

return
% ****************************************************************
```

Appendix 2 Matlab routine for 'tone control' evaluation

```
% Matlab routine for tone-control evaluation
% Tone control frequency response
% 04.08.13
close; home; clear; colordef white
fprintf('Tone-control frequency response\n')

% set ww = 1 to write Excel file of frequency responses
ww=0;

% frequency vector (1 Hz to 100 kHz, N points/dec)
N=1000; % frequency bins per decade
f=10.^(1:1/N:5); % frequency range 1 Hz to 100 kHz
dum=size(1:1/N:5);
NS=dum(2); clear dum; % number of frequency bins

% set data
fprintf('Load circuit and computation data\n')
maxg=5; % max gain/atten in dB
R1=5000;
RL=10000;
RH=10000;
C1=1*10^-6;
C2=.001*10^-6;

% display component values
mg=10^(round(maxg)/20);
R0=2*R1*(mg-1);
R2=2*R1; % identical lf and hf gain range
R0 %#ok<NOPTS>
RL %#ok<NOPTS>
RH %#ok<NOPTS>
R1 %#ok<NOPTS>
R2 %#ok<NOPTS>
C1 %#ok<NOPTS>
C2 %#ok<NOPTS>
```

```
% main program
fprintf('Evaluate and plot tone-control frequency response\n')
a=R0/(2*R1);
b=R0/R2;
c=RL/(8*R1);
p=0;
T0=i*2*pi*R0*C2;
T1=i*2*pi*R1*C1;
T2=i*2*pi*R2*C2;
T3=i*2*pi*RH*C2;
NP=(2*round(maxg)+1)^2;
tfdB=zeros(NP,NS);
for m1=-1:.2:1
for m2=-1:.2:1
p=p+1; % plot number
lambda1=a./((1+c*(1+m1)*(1-m1)+(c*(1+m1)*(1-m1)+.5)*f*T1));
lambda2=b*f*T2./(1+f*T2+0.25*T3*(1+m2)*(1-m2));
num=1+lambda1*(1-m1)/2+lambda2*(1-m2)/2;
den=1+lambda1*(1+m1)/2+lambda2*(1+m2)/2;
tfdB(p,1:NS)=20*log10(abs(num./den));

% plot tone-control frequency response
semilogx(f,tfdB(p,1:NS),'k')
hold on
end
end
title('Family of tone-control frequency responses')
xlabel('Frequency, Hz (logarithmic scale)')
ylabel('Response level, dB')
grid; pause(.5);

% write Excel file for frequency responses
if ww==1 % write Excel file
xlswrite('tone_control_response', (f;tfdB)');
end
pause; close; % close graphs

% end of program
fprintf('Program terminated\n')
return
% ******************************
```

Appendix 3 Matlab routine for parametric equalizer evaluation

```
% Parametric eq frequency response
% 04.08.13
close; home; clear; colordef white
fprintf('Parametric equalizer frequency response\n')

% set ww = 1 to write Excel file of frequency responses
ww=0;

% frequency vector (1 Hz to 100 kHz, N points/dec)
N=1000; % frequency bins per decade
f=10.^(1:1/N:5); % frequency range 10 Hz to 100 kHz
dum=size(1:1/N:5);
NS=dum(2); clear dum; % number of frequency bins

% set data
fprintf('Load circuit and computation data\n')
gmax(1)=15; % band 1 max boost/cut, dB
gmax(2)=15; % band 2 max boost/cut, dB
gmax(3)=15; % band 3 max boost/cut, dB
gmax(4)=15; % band 4 max boost/cut, dB
gmax(5)=15; % band 5 max boost/cut, dB
Q(1)=10; % band 1 Q-factor
Q(2)=1; % band 2 Q-factor
Q(3)=5; % band 3 Q-factor
Q(4)=3; % band 4 Q-factor
Q(5)=10; % band 5 Q-factor
fc(1)=15000; % band 1 centre frequency
fc(2)=5000; % band 2 centre frequency
fc(3)=1000; % band 3 centre frequency
fc(4)=250; % band 4 centre frequency
fc(5)=50; % band 5 centre frequency

% main program
fprintf('Evaluate and plot parametric equalizer frequency response\n')
q=1./Q; % 1/Q vector
gm=10.^(gmax(1:5)/20)-1; % gain vector
tfdB=zeros(31,NS); p=0;
for m=-1:1/15:1 % gain parameter
```

```
p=p+1;
num=zeros(1,NS); den=num;
for x=1:5;
num=num+.5*gm(x)*(1-m)*(i*q(x)*(f/fc(x)))./(1+i*q(x)*(f/fc(x))-(f/fc(x)).^2);
den=den+.5*gm(x)*(1+m)*(i*q(x)*(f/fc(x)))./(1+i*q(x)*(f/fc(x))-(f/fc(x)).^2);
end;
tfdB(p,1:NS)=20*log10(abs((1+num)./(1+den)));
semilogx(f,tfdB(p,1:NS),'k')
hold on
end
title('Parametric equalizer frequency response')
xlabel('Frequency, Hz')
ylabel('Response level, dB')
grid; pause(.5);

% write Excel file for frequency responses
if ww==1 % write Excel file
xlswrite('parametric_response', (f;tfdB)');
end
pause; close; % close graphs


% end of program
fprintf('Program terminated\n')
return
% **********************************
```

Optimising RIAA realisation

Douglas Self

RIAA equalisation is associated with moving-magnet (MM) pickups, but is no less necessary for moving-coil (MC) pickups. A very common arrangement is to have an MM input stage, with a suitable 47 kΩ input impedance, which can be switched between a directly-connected MM cartridge, and an MC cartridge followed by a flat low-noise gain stage. The noise conditions for the MM stage are very different in the two cases, but in either RIAA equalisation is required. This can be done in many ways, but some approaches can be discarded at once. Shunt-feedback RIAA stages for MM pickups are some 14 dB noisier than an equivalent series-feedback stage [1]; this is inherent because of the high inductance of MM cartridges, so we can lose that approach at once. Other methods split the various RIAA time-constant across separate stages, in various mixtures of passive and active circuits; this makes the calculations easy but uses more hardware and invariably involves compromises in headroom and noise performance.

If that sort of thing is the "brawn" hardware-heavy option, the "brains" option is to do all the RIAA equalisation in one stage, using series feedback. The "brains" part comes in because the interaction of the RIAA time-constants in a single feedback path makes the calculations for accurate RIAA much more difficult. Most of the hard work was done for us by Stanley Lipshitz, who in 1979 published comprehensive design equations for a large number of RIAA network configurations [2]. He deserves the thanks of us all. Nevertheless, these equations are pretty long and complicated, and adapting them to spreadsheet or MathCAD use is not a trivial business, so it is wise to assess which configurations are worth the effort before you undertake it.

At this point I am going to firmly state that the best way to implement RIAA equalisation is the conventional one-stage series-feedback method, using a gain of between +30 and +35 dB (at 1 kHz). Without going into all the detail of recorded levels and cartridge sensitivities, those gains will give more than adequate maximum input levels of 316 mVrms and 178 mVrms respectively. The output levels will be 158 and 281 mVrms respectively with a 5 mVrms input (1 kHz). The input stage will normally be followed by some form of controlled gain to get the signals up to the desired final level.

In this article RIAA accuracy is often worked out to a hundredth of a dB. That does not of course mean that such accuracy is essential for good listening, which is just as well as normal component tolerances would make it quite impossible. Instead the aim is to get the *nominal* results spot-on, so that whatever tolerances may do to us, we at least know we are aiming at the right values, and the practical results will be centred on them.

Equalisation and its discontents : the RIAA characteristic

The RIAA characteristic is probably the most complex standard equalisation curve in common use. **Figure 1** shows the response asymptotes of its main features. The most important three corners in its response curve are f3 at 50.05 Hz, f4 at 500.5 Hz, and f5 at 2.122 kHz, which are set by three time-constants of 3180 µ sec, 318 µ sec, and 75 µsec. To put it concisely, the f3 – f4 region allows for the constant-amplitude recording used over that range, while the plateau at f4 – f5 allows for the high-frequency recording pre-emphasis which reduces noise by de-emphasis at playback.

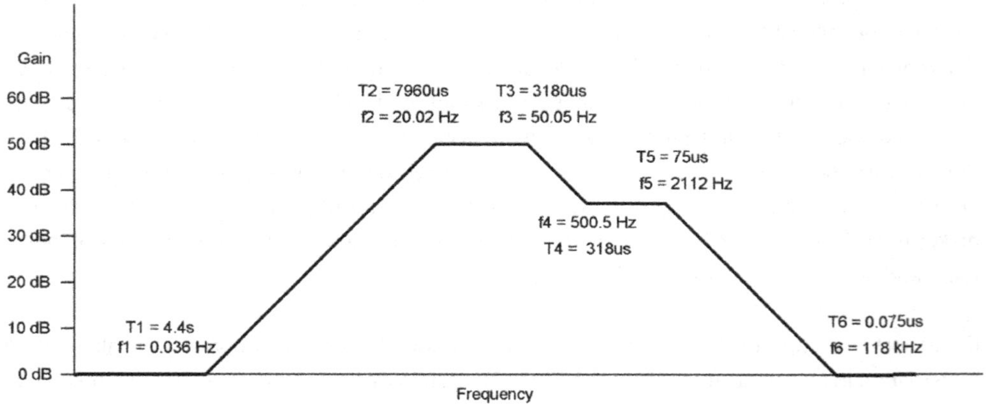

Figure 1 The practical response for series-feedback RIAA equalisation, including the IEC Amendment which gives an extra roll-off at f2 (20.02 Hz).

In Figure 1 there is a further response corner at f2 (20.02 Hz), corresponding to a time-constant of 7950 µs. This LF roll-off is called the "IEC Amendment" because it was added to the RIAA standard in 1976. It was intended to reduce the subsonic output, but its introduction is slightly mysterious. It was certainly not requested or wanted by either equipment manufacturers or their customers. Some manufacturers refused to implement it or just ignored it. It still attracts trenchant criticism today. The likeliest explanation for its introduction seems to be that various noise reduction systems (such as dbx), were then being promoted for use with vinyl and their operation was interfered with by subsonic disturbances (none of the systems caught on). In some preamps the IEC Amendment is switchable in/out.

The response in Figure 1 flattens out at f6, when the gain demanded by the RIAA network falls to unity, whereas to achieve the RIAA standard it ought to continue to drop at 6 dB/octave. This unwanted response corner is inherent in the series-feedback configuration, and is its only drawback. The gain in Figure 1 is +35 dB at 1 kHz, which puts f6 at 118 kHz, and this causes the gain to rise above the RIAA characteristic around 10 – 20 kHz; here it gives an excess gain of only 0.10 dB at 20 kHz, but the effect is more severe at lower gains because f6 occurs at lower frequencies. If the gain is reduced to +30 dB at 1 kHz to get more overload margin, the final zero f6 is now at only 66.4 kHz, introducing an excess gain at 20 kHz of 0.38 dB. This problem is completely cured by adding a lowpass time-constant after the stage; this is usually called an "HF correction pole" which means precisely the same thing. If the correct time-constant is used the gain and phase are corrected exactly.

The frequency f1 simply shows where the gain flattens out at the low-frequency end as once again it cannot fall below unity; it is usually of little or no importance.

A basic RIAA equalisation stage

Figure 2 shows a single-stage series-feedback MM preamp with a gain of +30 dB (1 kHz). It gives 158 mV out for 5 mV in (1 kHz), with an input overload level of 316 mV rms. This means an excellent overload margin of 36 dB, though more (variable) gain will be required later in most cases.

There are several different ways to arrange the resistors and capacitors in an RIAA network, all of which give identically exact equalisation when the correct component values are used. Of the four series RIAA configurations examined by Lipshitz, this is configuration A. It has the advantage that it is the simplest case when the Lipshitz equations are put into practice. Details that are essential for practical use, like input DC-blocking capacitors, DC drain resistors, and EMC/cartridge-loading ca-

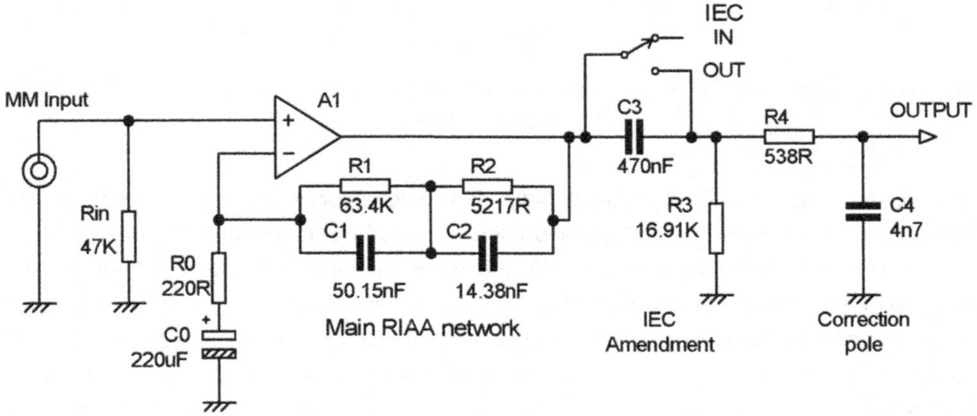

Figure 2 Series-feedback RIAA equalisation in configuration A, designed for 30.0 dB gain (1 kHz) which allows a maximum input of 316 mV rms (1 kHz). The switchable IEC Amendment is implemented by C3, R3. HF correction pole R4, C4 is added to keep RIAA accuracy within ± 0.1 dB 20 Hz to 20 kHz.

pacitors have been omitted for clarity; in later diagrams the 47 kΩ input loading resistor is also omitted. The overall impedance of the RIAA network has been kept reasonably low by making R0 220 Ω; further reduction has virtually no effect on the noise performance and results in large and expensive RIAA capacitors. No attempt is made at this point to deal with the awkward values of R1 and R2. The notation R0, C0, R1, C1, R2, C2 is as used by Lipshitz; C1 is always the larger of the two.

The unloved and unwanted IEC Amendment was no doubt intended to be implemented by restricting the value of C0; in this case the correct value of C0 would be 36.18 uF. This however leaves us wide open to significant errors due to the tolerance of an electrolytic capacitor. **Figure 3** shows that the gain will be +0.7 dB up at 20 Hz for a +20% C0, and -1.1 dB down at 20 Hz for a -20% C0. The possible effect of C0 is less than ±0.1 dB above 100 Hz, but this is obviously not a great way to make dependably accurate RIAA networks. It also requires non-standard values for either R0 or C0.

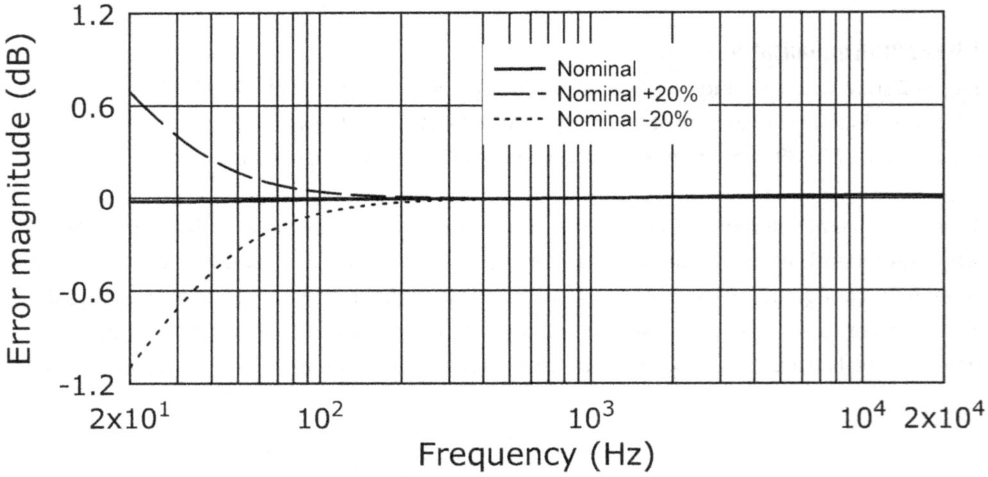

Figure 3 The effect of a ±20% tolerance for C0 when it is used to implement the IEC amendment in an amplifier; R0 = 220 Ω, gain +30 dB (1 kHz).

Instead C0 is increased to 220 uF so that its -3 dB roll-off does not occur until 3.29 Hz. Even this wide frequency spacing introduces an unwanted 0.128 dB loss at 20 Hz (assuming no IEC amendment is used), and perfectionists will want to use 470 uF here, which reduces the error to 0.06 dB. The IEC amendment is now implemented passively after the amplifier stage, by non-electrolytic capacitor C3 (which can have a tolerance of ±1 %) in conjunction with R3, giving the required -3 dB roll-off at f2 (20.02 Hz). It is easier to make it switchable in/out. In Figure 2 we use a standard E3 capacitor value for C3; 470 nF was chosen, and as usual an unhelpful resistor value results; in this case R3 = 16.91 kΩ. Using E24 resistors, this can be implemented exactly as 16 kΩ + 910 Ω, though it has to be said near-equal series or parallel resistor pairs would give more accuracy for a given tolerance; more on that later. The switch as shown may not be entirely click-free because of the offset voltage at A1 output, but that is not important as it will probably only be operated a few times in the life of the preamp, if indeed ever.

We made C0 220 uF, which will be a handily compact component. Because it is not infinitely large there is, as we saw, an error of -0.128 dB at 20 Hz. It is possible to compensate for this by tweaking the IEC amendment frequency f2. If R3 is changed from 16.91 kΩ to 17.4 kΩ the overall response is made accurate to ±0.005 dB. The compensation is not mathematically exact- there is a + 0.005 dB hump around 20 Hz- but I suggest it is good enough for most of us. This process gets the nominal response right but does not of course do anything to reduce the effects of the tolerance of C0; these are however small, around + 0.05 dB.

Since the gain is +30 dB (1 kHz) f6 is at 66.4 kHz and an HF correction pole is essential. This is implemented by R4, C4 in Figure 2. There is no significant interaction with the IEC amendment, so we have complete freedom in choosing C4 and use a standard E3 value and then get the pole frequency exactly right by using two resistors in series for R4- 470 Ω and 68 Ω. Since these components are only doing a little fine tuning at the top of the frequency range, the tolerance requirements are relaxed compared with the main RIAA network. The design considerations are a) that the resistive section R4 should be as low as possible in value to minimise Johnson noise, and on the other hand b) that the shunt capacitor C4 should not so large as to load the opamp output excessively at 20 kHz. It is assumed that an opamp such as the 5534 with good load-handling capability is used.

The IEC network is placed before the HF correction pole, as in Figure 2, so that R4 is not loaded by R3, which would cause a 0.3 dB loss, compromising noise or headroom slightly. As it is C3 is loaded by C4, but the loss is much smaller at 0.09 dB. It is assumed there is no significant external loading on the output at C4. Often the RIAA stage will be feeding the high-impedance input of a non-inverting gain stage, but if not buffering will be required so the two passive output networks behave as designed.

There is our basic RIAA stage. It works very well in both practice and theory, but is it the best solution possible? Here we set out on our voyage of optimisation.

Series-feedback RIAA-network configurations

Four possible configurations are described by Lipshitz in his classic paper. [2] These, using his component values, are shown in Figure 4; I have used the same identifying letters as Lipshitz. They are all accurate to within ±0.1 dB when implemented with a 5534 opamp, but in the case of configuration A the error is getting close to -0.1 dB at 20 Hz due to the relatively high closed-loop gain (+46.4 dB at 1 kHz) and the finite open-loop gain of the 5534. All have their RIAA networks at a relatively high impedance, and have relatively high gain and therefore a low maximum input. In each case the IEC amendment is implemented by the value of C0.

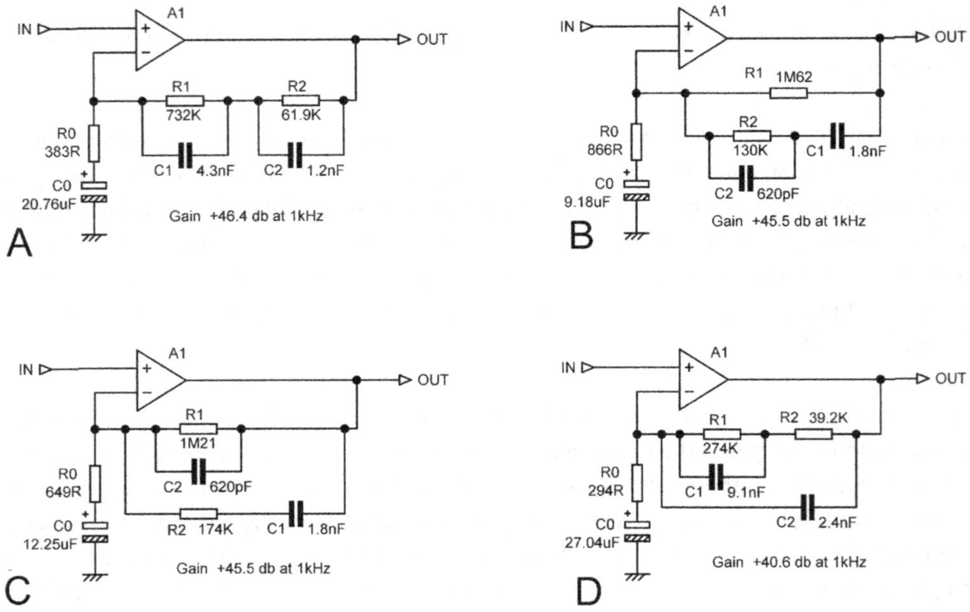

Figure 4 The four RIAA feedback configurations in the Lipshitz paper.

Burkhard Vogel in his weighty book *The Sound of Silence* describes configuration A as a Type-Eub network, configuration B a Fub-B network, and configuration C a Fub-A network [3]. He does not deal with configuration D.

Long ago I wrote a spreadsheet to implement the Lipshitz equations for configuration A simply because it was the easiest one to do. Repeating that for the remaining three configurations would be a fair amount of work, so the question arises, do any of the other three configurations have advantages over A that would make that work worthwhile?

Four things to examine suggested themselves:

1) Each configuration in Figure 4 contains two capacitors, a large C1 and a small C2 that set the RIAA response. If they are close-tolerance (to get accurate RIAA) and non-polyester (to prevent capacitor distortion) then they will be expensive, so if there is a configuration that makes the large capacitor smaller, even if it is at the expense of making the small capacitor bigger, it is well worth pursuing. That large capacitor is almost certainly the most expensive component in the RIAA MM amplifier by a long way.

2) I started by assuming that the signal voltages across each capacitor would be different. If polyester capacitors must be used for cost reasons, then if there is a configuration that puts less voltage

across a capacitor then that capacitor will generate less distortion. Capacitor distortion at least triples, and may quadruple, as the voltage across it doubles, so choosing the configuration that minimises the voltage might be worthwhile. However, a good deal of simulation tells us that there is actually little difference between the four configurations as regards the magnitude of the signal voltage across the capacitors. Not a helpful result or an interesting result, but sometimes things just are that way.

3) If we assume a certain amount of non-linearity in one or both of the RIAA capacitors, are the configurations different in their sensitivity to that non-linearity? In other words, how much distortion will appear at the output? This could be resolved in SPICE simulation, by using non-linear capacitor models constructed with Analog Behavioural Models, but it would be a lot of work, and since the emphasis of this article is on high quality, where we can presumably afford a polypropylene capacitor or two, I have put that one on the back-burner, though I hope it will not fall completely off the back of the cooker.

4) My implementation of the Lipshitz equations starts off with the value of R0. Combined with the desired gain, it sets the impedance level of the whole RIAA network. A low impedance level reduces noise, but increases the size and cost of the precision capacitors C1 and C2. So just how low should R0 be for the best performance? This issue is examined at the end of this article.

That leaves us with 1) and 4), and in these cases we *do* get some useful results. Now read on...

The RIAA configurations compared for capacitor cost

We need to put the four configurations A, B, C, and D into a common form where they can be directly compared. In Figure 4 they are working at different impedance levels, as shown by the differing values of R0, and at different gains. To make the impedance levels the same we scale all the RIAA component values to make R0 exactly 200 Ω, as in **Figure 5**. C0 is then 39.75 uF in each case, and implements the IEC amendment. The impedance scaling does not affect the gain or the RIAA accuracy; this was checked by simulation for each configuration. The new values for C1 and C2 are shown in **Table 1**.

Configuration	Gain at 1 kHz	Large cap C1	Small cap C2	C1/C2 ratio
A	+46.4 dB	8.235 nF	2.298 nF	3.583
B	+45.5 dB	7.794 nF	2.685 nF	2.903
C	+45.5 dB	5.841 nF	2.012 nF	2.903
D	+40.6 dB	13.38 nF	3.528 nF	3.791

Table 1 The values of C1 and C2 in Figure 5, with networks scaled so R0 =200 Ω in each case.

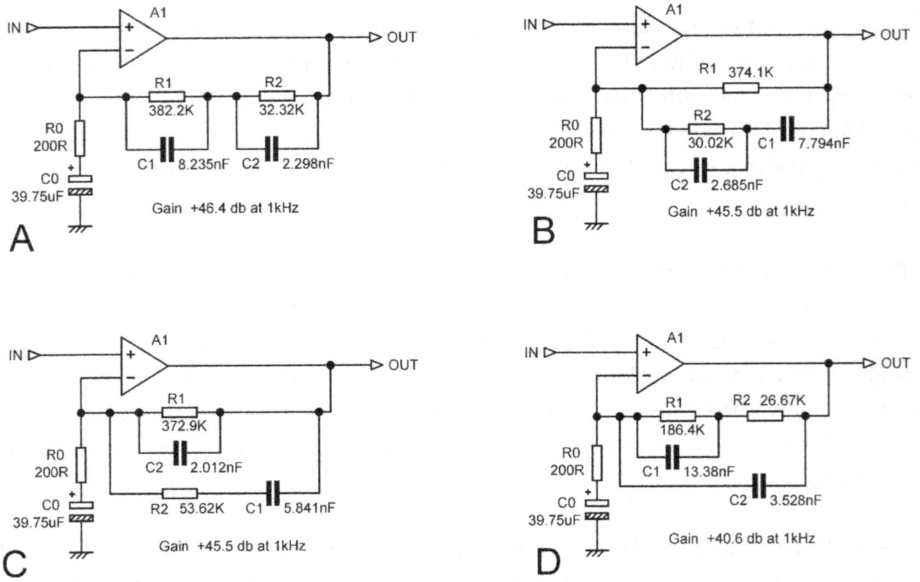

Figure 5 The four RIAA feedback configurations, with component values scaled so that R0 = 200 Ω in each case.

Since the gains of configurations A, B, and C are nearly the same, we can roughly compare the values for C1, and it's already starting to look as if A might in fact be the worst case for capacitor size. To be quite sure about this we can alter the gain of A to be exactly the same as B and C at +45.5 dB, which is easy as we already have the spreadsheet for it. We don't have the tool for D. Configuration D has significantly bigger capacitors than A, B, and C because it has about half the gain but the same value of R0.

We can change the gain of D by simply by scaling the RIAA network, with R0 kept constant. This will not be very accurate but should be good enough for us to compare the configurations. We have to increase the gain of D by 4.948 dB, or 1.767 times, so we divide capacitors C1 and C2 by this factor,

Configu-ration	Gain at 1 kHz	Large cap C1	Small cap C2	C1/C2 ratio
A	+45.5 dB	9.197 nF	2.568 nF	3.581
B	+45.5 dB	7.794 nF	2.685 nF	2.903
C	+45.5 dB	5.841 nF	2.012 nF	2.903
D	+45.5 dB	7.567 nF	1.996 nF	3.791

Table 2 The values of C1 and C2 in Figure 6, after scaling so R0 =200 Ω and gain =+45.5 dB (1 kHz) for all four configurations

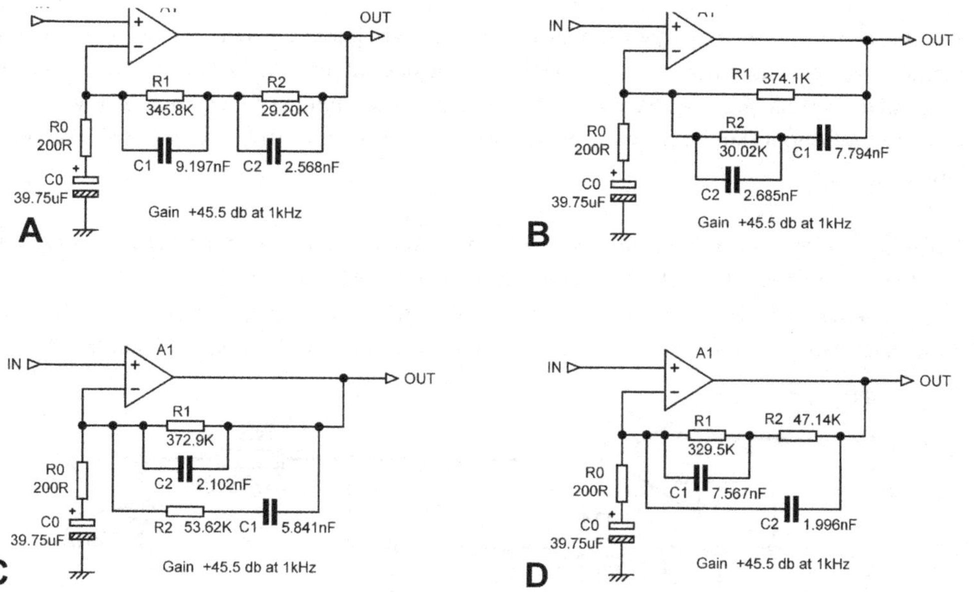

Figure 6 The RIAA feedback configurations, with component values scaled so that R0 = 200 Ω and the gain is +45.5 dB at 1 kHz in each case. Note that the RIAA response of D is here not completely accurate. The high gain means that the maximum input in each case is only 53 mV rms (1 kHz) which is not in general adequate.

and multiply resistors R1 and R2 by it. This gives the values shown in **Figure 6** and **Table 2**. The C1/C2 ratios are naturally unchanged.
After this gain alteration, configuration D is definitely less accurate than before, but is still almost within a ±0.1 dB error band, only exceeding this (and not by much) over small frequency ranges. This is good enough to allow proper assessment of each configuration.

From Figure 6 and Table 2, we can see that C1 in configuration C is only 63% the size of C1 in configuration A. I expected this, but was afraid it might be accompanied by an increase in C2 in configuration C, but this is also smaller at 78% of C2 in A. I have to ruefully admit that configuration A (which I have been using for years) makes the least efficient use of its capacitors, since they are essentially in series, reducing the effective value of both of them. Configurations B and D have intermediate values for C1, but of these two D has a significantly smaller C2.

Therefore configuration C would appear to be the optimal solution in terms of capacitor size, and hence cost and PCB area used. B and D appear to have no special advantages. It was now time to put together a software tool for the Lipshitz equations of C, so it could be designed accurately for gains other than +45.5 dB (1 kHz). As anticipated it was somewhat more difficult than it had been for configuration A, but no big deal.

We have already noted that a gain as high as +45.5 dB (1 kHz) gives an unacceptably poor overload margin. It has only used here so far because it was the gain adopted in the Lipshitz paper. If we assume an opamp can provide 10 Vrms out, then the maximum input at 1 kHz is only 53 mV rms. The gain of an MM input stage should not, in my opinion, much exceed 30 dB (1 kHz) See the earlier example in Figure 2.

My Precision Preamplifier design [4] has an MM stage gain of +29 dB (1 kHz), permitting a maximum input of 354 mV rms (1 kHz). The more recent Elektor Preamplifier 2012 [5] has an MM stage gain of +30 dB (1 kHz), allowing a maximum input of 316 mV rms; it is followed by a flat switched-gain stage which allows for the large range in MC cartridge sensitivity. Both designs use configuration A.

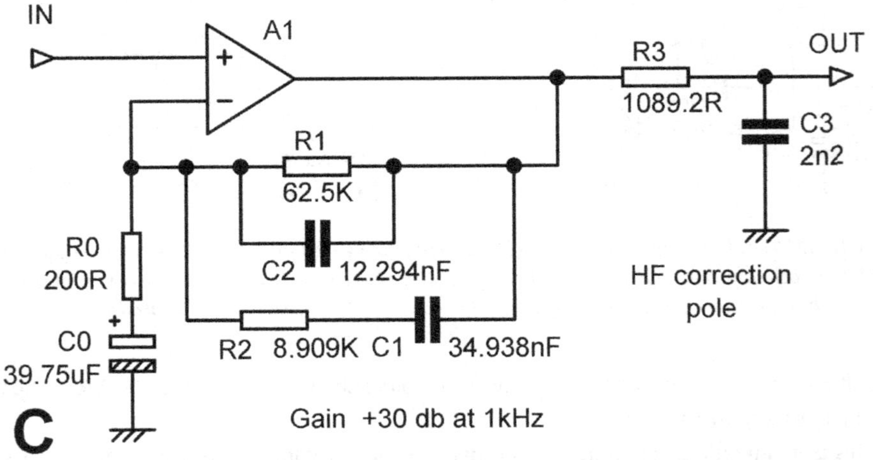

Figure 7 Configuration C with values calculated from the Lipshitz equations to give +30.0 dB gain at 1 kHz, and an accurate RIAA response within ±0.01 dB; the lower gain now requires HF correction pole R3, C3 to maintain accuracy at the top of the audio band.

A preamp with a gain of exactly +30 dB (1 kHz) using configuration C was therefore designed with the new software tool, and there it is in **Figure 7**. It has an RIAA response, including the IEC amendment that is accurate to within ±0.01 dB from 20 Hz to 20 kHz (assuming C0 is exactly correct in value). The relatively low gain means that an HF correction pole is required to maintain accuracy at the top of the audio band, and this is implemented by R3 and C3. If it was omitted the response would be 0.1dB high at 10 kHz, and 0.37 dB high at 20 kHz. We use the E6 value of 2n2 for capacitor C3, so R3 is a non-preferred value.

In Figure 7, and in the examples that follow, I have implemented the IEC amendment by using the appropriate value for C0, rather than by adding an extra time-constant after the amplifier as in Figure 2. We have already seen that using C0 to do this is not the best method, but I have stuck with it here as it is instructive how the correct value of C0 changes as other alterations are made to the RIAA network.

Making C1 a single E6 capacitor

So far we have optimised our RIAA network by switching to configuration C. A further stage of refinement is possible after minimising C1 and C2. There's nothing special about the value of R0 at 200 Ω (apart from the bare fact that it's an E24 value), it just needs to be suitably low for a good noise performance. We can therefore scale the RIAA components again, not to make R0 a specific value, but to make at least one of the capacitors a preferred value, the larger one being the obvious candidate. Compared with the possible savings on expensive capacitors here, the cost of making up a non-preferred value for R0 is negligible.

In Figure 7, C1, at 34.9 nF, is suggestively close to 33nF. If we twiddle the new software tool for configuration C so that C1 is exactly 33 nF, we get the circuit in **Figure 8**. R0 has only increased by 6% to 212 Ω, and so the effect on noise performance will be negligible; I calculate it as 0.02 dB worse (with MM cartridge input load). All the other values in the RIAA feedback network have likewise altered by about 6%, including C0, but the HF correction pole is unchanged; that would only need to be altered if we altered the gain of the stage. Like Figure 7, the RIAA accuracy of this circuit is well within ±0.01 dB from 20 Hz to 20 kHz when a 5534 is used.

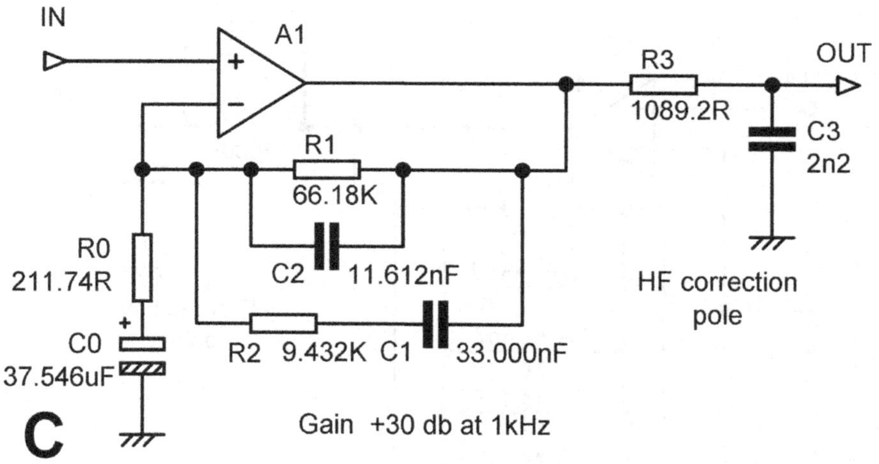

Figure 8 Configuration C from Figure 7 with Ro tweaked to make C1 exactly the E6 preferred value of 33.000 nF. Gain is still 30.0 dB at 1 kHz, and RIAA accuracy within ±0.01 dB. The HF correction pole R3, C3 is unchanged.

The circuit of Figure 8 now has two preferred-value capacitors, C1 and C3, but that is the best we can do. All the other values are, as expected, thoroughly awkward. In my book *Active Crossover Design* [6] I describe how to make up arbitrary resistor values by paralleling two or more resistors, and how the tolerance errors partly cancel. This also applies to capacitors. The optimal way to do this is with components of as near-equal values as you can manage. If the values are exactly equal, the combined accuracy is √2 time better than the individual components; the further away from equality, the less the improvement. Nothing more esoteric than the E24 series is assumed for the resistors. The optimal parallel pairs were selected in a jiffy using a specially written software tool. The parallel combination for C2, using E6 values, was done by manual bodging, though as you can see in **Table 3**, we have been rather lucky with how the values work out, with three parallel capacitors getting us very close to the desired value. **Figure 9** shows the resulting circuit.

Compo-nent	Desired value	Actual value	Parallel part A	Parallel part B	Parallel part C	Error
R0	211.74 Ω	211.03 Ω	360 Ω	510 Ω	-	-0.33%
R1	66.18 kΩ	65.982 kΩ	91 k Ω	240 k Ω	-	-0.30%
C1	33 nF	33 nF	33nF	-	-	0%
R2	9.432 kΩ	9.474 kΩ	18 kΩ	20 kΩ	-	0.44%
C2	11.612 nF	11.60 nF	4n7	4n7	2n2	-0.10%
R3	1089.2 Ω	1090.9Ω	2 kΩ	2.4 kΩ	-	0.16%

Table 3 Approximating the exact values of Figure 8 with paralleled components, giving Figure 9.

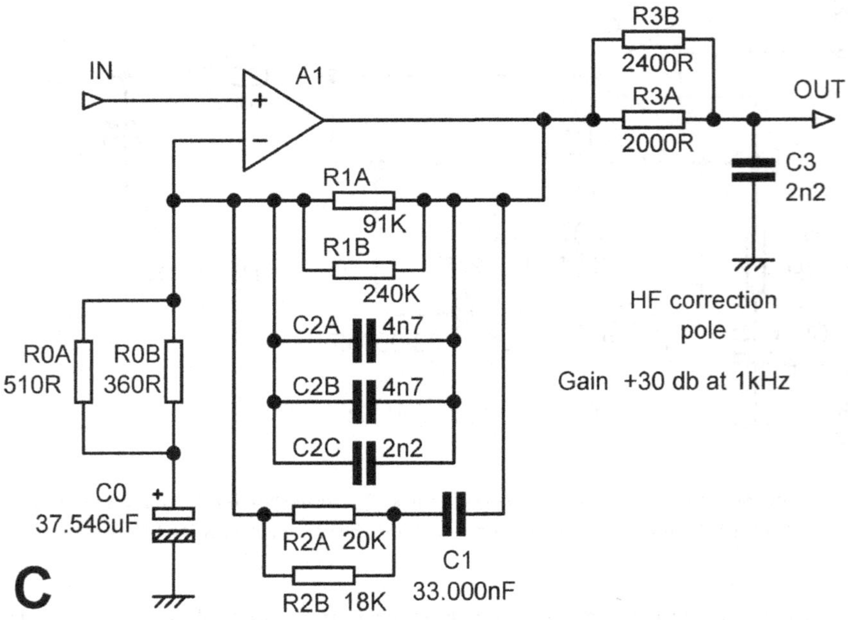

Figure 9 Configuration C from Figure 8 with the resistors made up of optimal parallel pairs to closely approach the correct value. C2 is made up of three parts. Gain +30.05 dB at 1 kHz, RIAA accuracy is worsened but still within ±0.048 dB.

When selecting the parallel resistor pairs, the criterion I used was that the error in the nominal value should be less than half of the component tolerance, assumed to be ±1 %. R2 only just squeaks in at +0.44 %, but its near-equal values give almost all of the √2 improvement. Don't forget that in Table 3 we are dealing with *nominal* values, and the % error in the nominal value shown in the rightmost column has nothing to do with the component tolerances.

Naturally the small errors in nominal value seen in the rightmost column of Table 3 have their effect. **Figure 10** shows that the RIAA response now has a gentle peak of +0.048 dB at 1 kHz, which is by coincidence the worst-case frequency. Above that, the error slowly declines to +0.031 dB at 20 kHz. Most of this deviation is caused by the +0.44 % error in the nominal value of R2 (see Table 3).

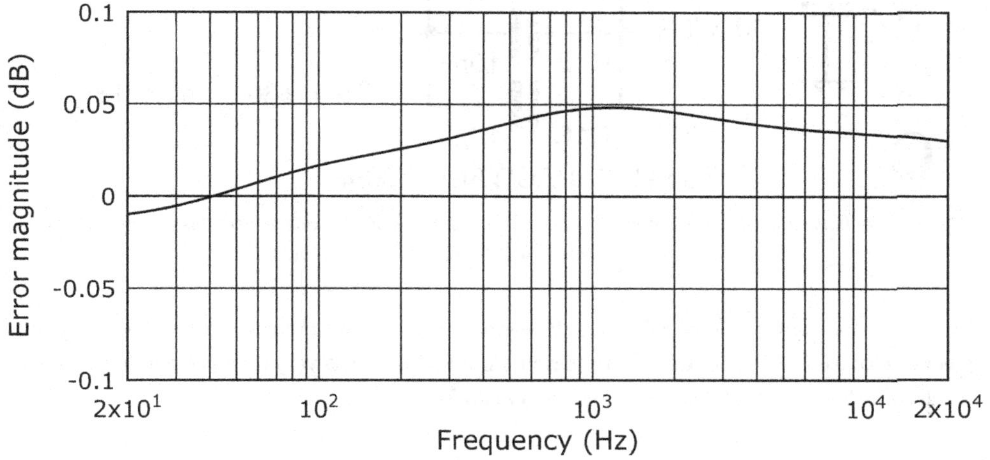

Figure 10 The RIAA accuracy of Figure 9. Gain is 30.05 dB at 1 kHz, and RIAA error reaches a maximum of +0.048 dB midband.

C1 made up of multiple 10nF capacitors

We have just scaled the RIAA network so that the major capacitor C1 is a single preferred value; at 33nF it will probably be a rather expensive 1% polypropylene. RIAA optimisation can however be tackled in another way, depending on component availability. In many polystyrene capacitor ranges 10 nF is the highest value that can be obtained with a tolerance of 1%. Paralleling several of these is a good deal more cost-effective than a single 1% polypropylene part. To exploit this we need to redesign the circuit of Figure 8 so that C1 is either exactly 30 nF or exactly 40 nF. **Figure 11** shows the result for C1 = 30 nF. Ro has now increased to 233 Ω, which I calculate will degrade the noise by only 0.06 dB compared with R0 = 200 Ω (with MM cartridge input load).

The 40 nF version costs a bit more but gives a total capacitance that is twice as accurate as one capacitor (as √4 = 2), while the 30 nF version only improves accuracy by √3 (= 1.73) times. **Figure 13**

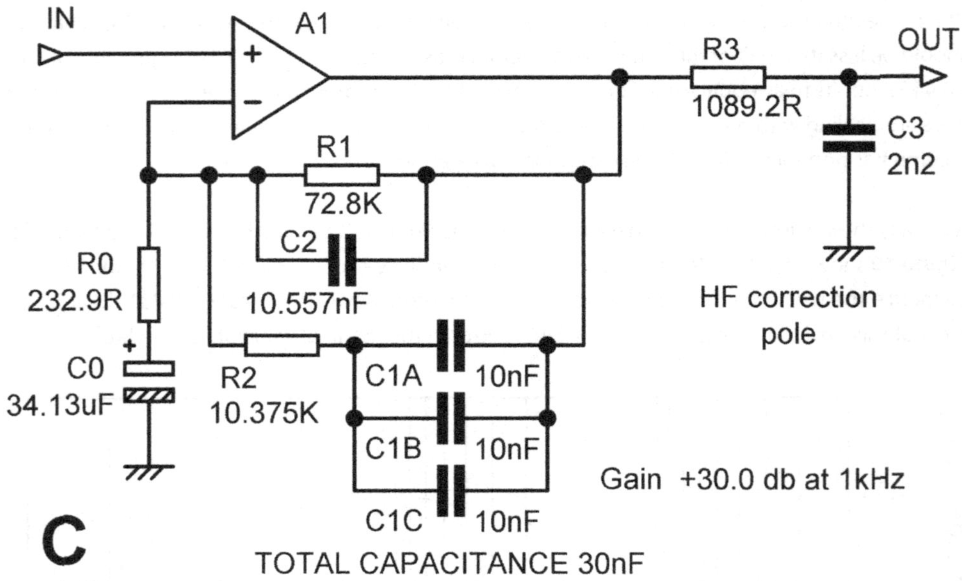

Figure 11 Configuration C from Figure 8 redesigned so that C1 is now 30 nF, made up with three paralleled 10 nF capacitors. Gain +30.0 dB at 1 kHz, RIAA accuracy is within ±0.01 dB.

shows the result for C1 = 40 nF. R0 has now dropped to 175 Ω, which gives a calculated noise improvement of 0.04 dB. Not to belabour the point, but R0 has only a very minor effect on the noise-we will come back to that later. C0 is now larger at 45.51uF.

Since in each case the gain is unchanged the values for the HF correction pole R3, C3 are also unchanged.

As for the previous example with C1= 33nF, the awkward resistor values are made up with optimally-selected parallel pairs. The results of this process for C1 = 30nF are shown in **Table 4** and **Figure 12**. In this case we have been unlucky with the value of C2, which needs to be trimmed with a 120 pF capacitor to meet the criterion that the error in the nominal value will not exceed half the component tolerance.

The same process can be applied to the 4 x 10 nF version in **Figure 13**, giving the component values in **Table 5** and in **Figure 14**.

This time we are much luckier with the value of C2; three 4n7 capacitors in parallel give almost exactly the required value. On the other hand we are very unlucky with R0, where 180 Ω in parallel with 6.2 kΩ is the most "equal" solution falling within the error criterion.

Both my Precision Preamplifier '96 [4] and the more recent Elektor Preamplifier 2012 [5] have MM stage gains close to +30 dB (1 kHz) like the examples above, but both use configuration A, and five

Compo-nent	Desired value	Actual value	Parallel part A	Parallel part B	Parallel part C	Error
R0	232.9 Ω	233.3 Ω	430 Ω	510 Ω	-	0.17%
R1	72.80 kΩ	72.97 kΩ	100 k Ω	270 k Ω	-	0.24%
C1	30 nF	30 nF	10 nF	10 nF	10 nF	0%
R2	10.375kΩ	10.359 kΩ	13 kΩ	51 kΩ	-	-0.15%
C2	10.557 nF	11.60 nF	4n7	4n7	1nF + 120 pF	0.34%
R3	1089.2 Ω	1090.9Ω	2 kΩ	2.4 kΩ	-	0.16%

Table 4 Approximation to the exact values in Figure 11 by using parallel components, giving Figure 12.

paralleled 10 nF capacitors are required for C1. A third existing design using configuration A was modified to configuration C as in Figure 12, and thoroughly tested with an AP SYS-2702. The RIAA was exactly correct within the limits of measurement and the noise performance was unchanged. It gave a hefty parts-cost saving of some £2 on the product concerned.

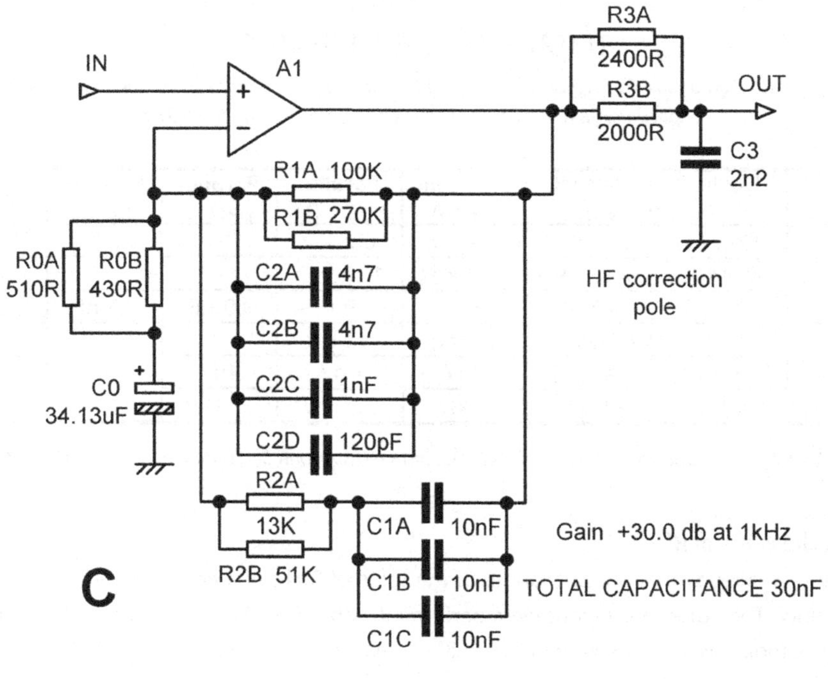

Figure 12 Configuration C from Figure 11 with resistors made up of optimal parallel pairs, and C2 made up of four parts. Gain +30.0 dB at 1 kHz, RIAA accuracy is within ±0.01 dB.

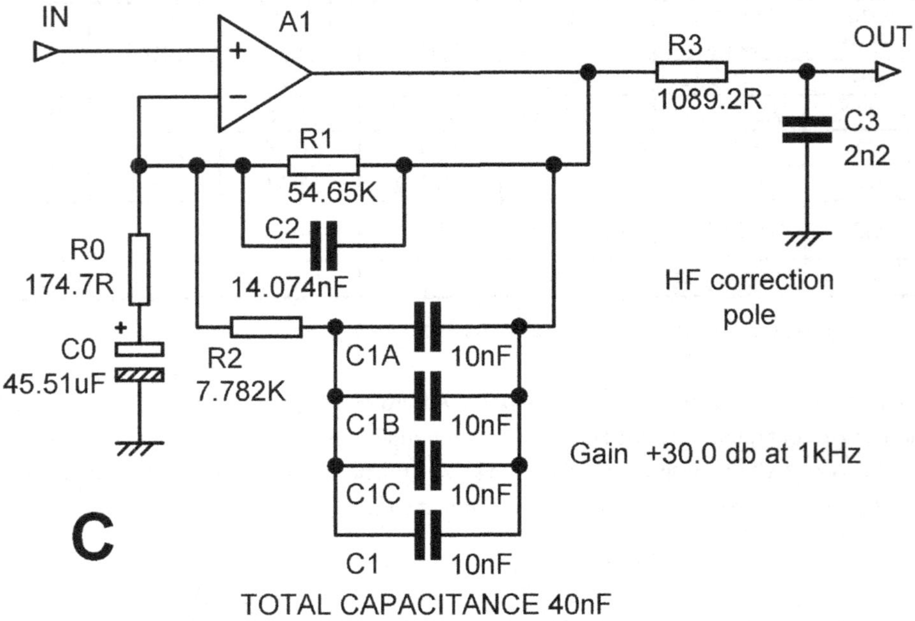

Figure 13 Configuration C from Figure 8 redesigned so that C1 is 40 nF, made up with four paralleled 10 nF capacitors. Gain +30.0 dB at 1 kHz, RIAA accuracy is within ±0.01 dB.

Component	Desired value	Actual value	Parallel part A	Parallel part B	Parallel part C	Parallel part D	Error
R0	174.7 Ω	174.9 Ω	180 Ω	6.2 kΩ	-	-	0.10%
R1	54.44 kΩ	54.54 kΩ	100 k Ω	120 k Ω	-	-	0.41%
C1	40 nF	40 nF	10 nF	10 nF	10 nF	10 nF	0%
R2	7.821kΩ	7.765 kΩ	12 kΩ	22 kΩ	-	-	-0.22%
C2	14.074 nF	14.1 nF	4n7	4n7	4n7	-	0.18%
R3	1089.2 Ω	1090.9Ω	2 kΩ	2.4 kΩ	-	-	0.16%

Table 5 Approximation to the exact values in Figure 13 by using parallel components, giving Figure 14.

Noise & the value of R0

We have seen already that minor changes in the value of R0 have a negligible effect on the noise performance. The noise calculations were performed with MAGNOISE2, an upgraded version of my MAGNOISE tool. This gives results within 0.2 dB of measurements and I believe it trustworthy.

In my published designs, R0 has gone from 470 Ω in 1979 [7] to 330 Ω in 1983 [8] and then 220 Ω in 1996 [4] as part of a general drive towards low noise by low-impedance design. I considered 220 Ω about as low as could be achieved while using a reasonable number (five) of paralleled 10 nF polystyrene capacitors for C1.

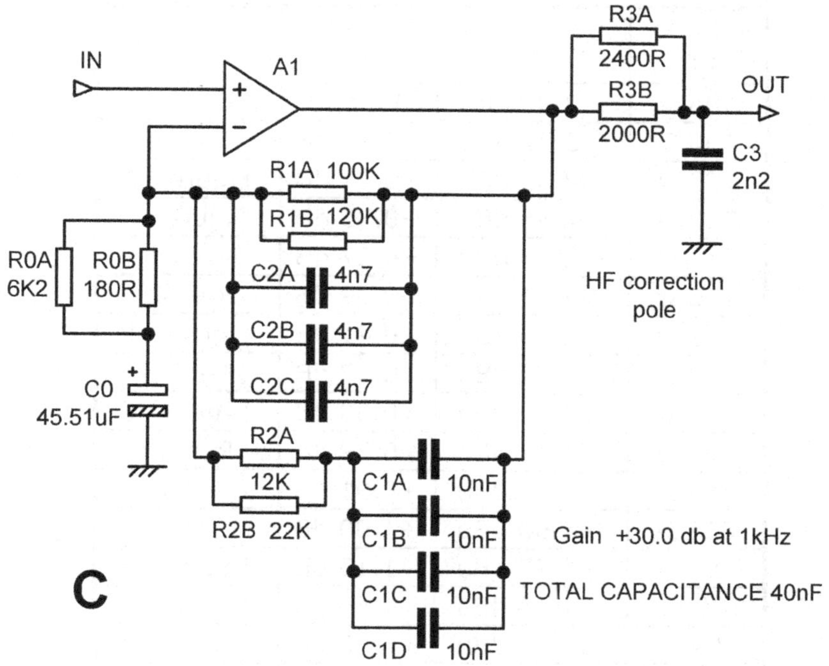

Figure 14 Configuration C from Figure 13 with resistors made up of optimal parallel pairs. C2 is made up of three parts. Gain +30.0 dB at 1 kHz, RIAA accuracy is within ±0.01 dB.

In MAGNOISE2, any source of noise can be switched off so its contribution can be assessed. R0 produces noise in two ways; its own Johnson noise, and the voltage produced by the opamp current noise flowing through it. At these impedances, the latter is negligible. **Table 6** shows the results for 30 dB gain (1 kHz) using a 5534AN opamp and cartridge parameters of R = 610 Ω, L = 470 mH, for the values of R0 used in the discussions above, plus some more. The noise results are shown to two decimal places simply so that the small changes are visible; in practice they will be swamped by variations in opamp noise. R0 has only a minor effect; most of the noise comes from the Johnson noise of the cartridge resistance and the 47 kΩ input load, the opamp voltage noise, and the opamp current noise at the non-inverting input.
The first thing we see is that with R0 at my usual value of 220 Ω, we are only 0.39 dB noisier than some impossible circuit in which R0 did not exist at all. That is not an audible difference, and suggests that if we can accept just a little more noise we would be able to increase the impedance of the RIAA network and further reduce the amount of expensive capacitance required.

Looking at configuration C in Figure 11, with R0 = 233 Ω, we can redesign it, using the new software tool, so that C1 = 2 x 10 nF. Running the numbers, we get R0 = 349 Ω, as in Table 7, which Table 6 shows will be noisier than Figure 11 by only 0.18 dB. In many cases this will be an acceptable cost/performance trade-off. If we were happy with a 1.0 dB noise increase, R0 can go as high as 611 Ω.

Value of R0	Noise out	Noise ref R0 = 0 Ω	Where used
0 Ω	-92.90 dBu	0.00 dB	
100 Ω	-92.72 dBu	+0.18 dB	
175 Ω	-92.59 dBu	+0.31 dB	Figure 13
200 Ω	-92.55 dBu	+0.35 dB	Figure 7
212 Ω	-92.53 dBu	+0.37 dB	Figure 8
220 Ω	-92.51 dBu	+0.39 dB	Ref [4]
233 Ω	-92.49 dBu	+0.41 dB	Figure 11
330 Ω	-92.33 dBu	+0.57 dB	Ref [8]
349 Ω	-92.30 dBu	+0.60 dB	Table 7 C
470 Ω	-92.11 dBu	+0.79 dB	Ref [7]
552 Ω	-91.99 dBu	+0.91 dB	Table 7 A
611 Ω	-91.90 dBu	+1.00 dB	
699 Ω	-91.77 dBu	+1.13 dB	Table 8 C
1000 Ω	-91.36 dBu	+1.54 dB	
1104 Ω	-91.22 dBu	+1.68 dB	Table 8 A
2000 Ω	-90.20 dBu	+2.70 dB	
2310 Ω	-89.90 dBu	+3.00 dB	

Table 6 Calculated noise output for different values of R0 (22 – 22 kHz bandwidth, with RIAA), unweighted.

For comparison Table 7 also shows the component values for configuration A, with C1 constrained to be 20 nF. Because of its inefficient use of capacitance, this requires R0 to be a good deal higher at 552 Ω, and the noise penalty referred to R0 = 233 Ω is 0.50 dB rather than 0.19 dB. This is rather less acceptable and confirms that configuration C is the best. Table 7 was checked by simulation and the RIAA accuracy is within ±0.01 dB from 20 Hz to 20 kHz.

Configu-ration	R0	C0	C1	C2	R1	R2
C	349.36 Ω	22.76 uF	20 nF	7.038 nF	108.87 kΩ	15.562 kΩ
A	552.24 Ω	14.39 uF	20 nF	5.727 nF	158.99 kΩ	13.094 kΩ

Table 7 Calculated RIAA components with C1 constrained to be 20 nF. IEC amendment implemented by C0.

Let's push this a little further and assume we want C1 to be a single 10 nF 1% capacitor for real economy. What is the noise penalty? The software tools give us **Table 8** for configurations C & A, and the noise penalties are inevitably larger, though perhaps not by as much as you would expect for fairly radical changes in the impedance of the RIAA network. Table 8 was checked by simulation and the RIAA accuracy is within ±0.01 dB from 20 Hz to 20 kHz.

Configu-ration	R0	C0	C1	C2	R1	R2	Noise ref R0=233 Ω
C	698.7 Ω	11.37 uF	10 nF	3.519 nF	217.74 kΩ	31.126 kΩ	+0.72 dB
A	1104.4 Ω	7.198 uF	10 nF	2.864 nF	317.99 kΩ	26.189 kΩ	+1.27 dB

Table 8 Calculated RIAA components with C1 constrained to be 10 nF; IEC amendment implemented by C0.

All the noise calculations above are for a preamplifier driven by an MM cartridge with the parameters given above. As we noted at the start, the MM RIAA stage may also be driven from an MC head-amp. The noise conditions for the opamp are quite different, as it is now fed from a very low impedance, probably dominated by a series resistor in series with the MC amp output to give stability against stray capacitances. My current MC head-amp design has an EIN of -141.5 dBu with a 3.3 Ω input source resistance [5]. Its gain is +30 dB, so the noise at its output is -111.5 dBu. The series output resistor is 47 Ω and its Johnson noise at -135 dBu is negligible, and so is the effect of the RIAA stage input noise flowing in it. The noise at the output of the RIAA stage is then -85.7 dBu, which is considerably higher than any of the figures in Table 6. In this application the value of R0 is uncritical.

Conclusions

Our investigations have shown that there are very real differences in how efficiently the various RIAA networks use their capacitors, and it looks clear that using configuration C rather than configuration A will cut the cost of the expensive capacitors C1 and C2 in an MM stage by 36% and 19% respectively, which I suggest is both a new result and well worth having. From there we went on to find that different constraints on capacitor availability lead to different optimal solutions for configuration C.

We have also demonstrated that accepting very small amounts of extra noise allows us to increase the impedance of the RIAA network substantially, further reducing the capacitance required and saving some more of our hard-earned money.

I hope you will forgive me for not making public the software tools mentioned in this article. They are part of my stock-in-trade as a consultant engineer, and I have invested significant time in their development.

References

1] Self, Douglas, "Small Signal Audio Design" Focal Press 2010, p171; ISBN 978-0-240-52177-0

2] Lipshitz, S P, "On RIAA Equalisation Networks" *J. Audio Eng Soc,* June 1979, p458 onwards.

3] Vogel, Burkhard, *The Sound of Silence,* Springer 2nd edition 2011, p523; ISBN 978-3-642-19773-d

4] Self, Douglas, "Precision Preamplifier 96" *Electronics World* July/August and September 1996

5] Self, Douglas, "Elektor Preamplifier 2012" *Elektor* April, May, June 2012

6] Self, Douglas, "Active Crossover Design" Focal Press, 2011, pp342-350, ISBN 978-0-240-81738-5

7] Self, Douglas, "High-Performance Preamplifier", Wireless World, Feb 1979, p40

8] Self, Douglas, "Precision Preamplifier", Wireless World, Oct 1983, p31

Towards Advanced Audio Power Amplifiers
Part 1 – on feedback, stability and minimum-phase

David Zan

Introduction

The previous article introduced the idea that amplifier advances centre around improvements in feedback.
This article is focussed on how to optimise the feedback properties of amplifiers.
The ideas will be explained by specific examples of advanced feedback techniques but the plan is not just to analyse particular feedback implementations.
The ultimate objective is a systematic way to find the optimum of all possible linear feedback interconnections, for a specified amplifier.

First it is helpful to expand and deepen some of the ideas from the first article and this can also act as a review and lead-in for any reader who missed the first part.
A deeper comprehension is important for feedback theory because many of the ideas are initially quite counter intuitive. Indeed the very first patent application for a feedback-stabilised amplifier was delayed because apparently the claims seemed nonsensical to the patent examiners.
Some of these misconceptions still turn up in audio discussions after almost a century.
Many audio myths have some kernel of sense, that's what makes them spread like a virus, or meme to use the term Richard Dawkins coined, so some will be discussed in the article as they arise.

The first and most fundamental point to establish is that increased feedback is in fact desirable.
Bruno Putzeys' article on this topic was published in Linear Audio Volume 1 and has rapidly become a classic, so the case for more feedback won't be repeated here [1].

Stability

The first analysis of feedback, by H.S. Black, made it clear that better performance could result from increased feedback [2]. But experiments, even prior to the theory, also made it clear that feedback could not be increased indefinitely. As the feedback is increased an amplifier eventually becomes unstable.
So the key problem is to understand exactly under what conditions a feedback system is stable.
The correct answer actually has some counter-intuitive solutions; systems can be stable even when common sense would imply the opposite.

This problem had already been studied a little for mechanical control systems prior to electronic amplifiers. But the keystone of all modern stability theory was developed by Harry Nyquist and published in the Bell System Technical Journal in 1932 [3]. The work was done so quickly that it was actually printed before the publication on feedback itself, delayed as mentioned earlier. But despite the speed of the work the paper is remarkably fully developed, in fact it is broader than is needed to analyse advanced audio amplifiers or even active feedback. It extends to cover systems that are naturally unstable without feedback, such as modern combat aircraft control systems and similar. This type of instability never occurs in any practical audio amplifier so the extra complications will be mostly skipped over here, in the interest of clarity.

Nyquist examined stability with the mathematical tools of complex analysis and explained the results with plots on the complex plane. This complex plane is used to plot a quantity with both amplitude and phase, and in this context is often called the "S plane" where "S" is the complex frequency parameter.

A Nyquist plot is the gain around the loop as a function of frequency, shown in linear coordinates (figure 1). Later developments have almost overshadowed the Nyquist plot but it remains available in most simulators. In LTspice it can be selected in the options for the left vertical axis.

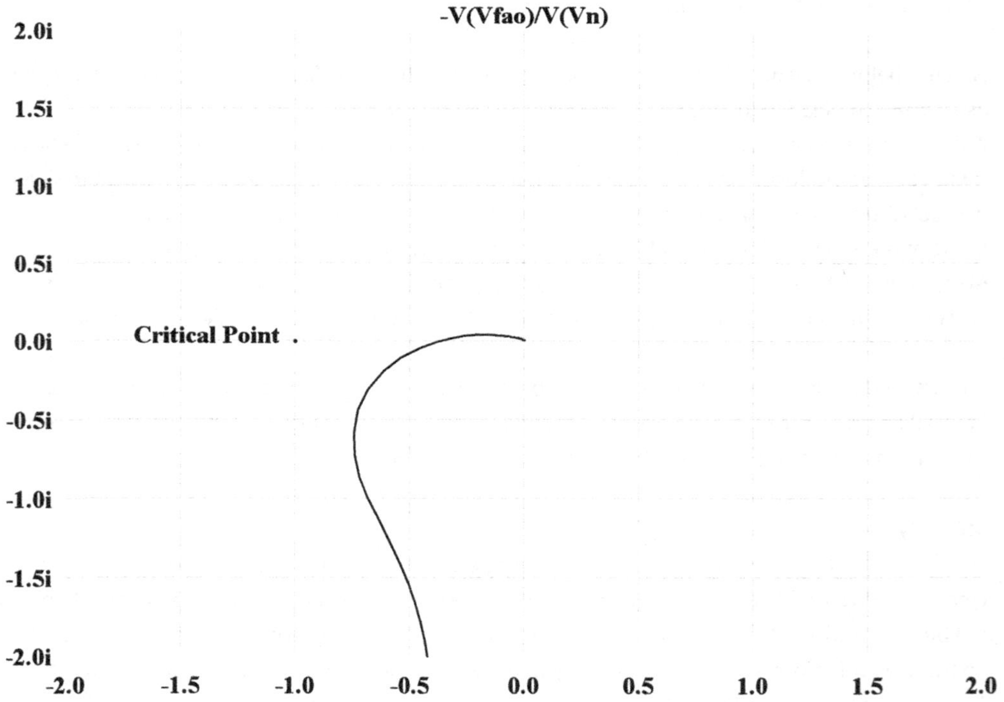

Figure 1 Nyquist plot for a typical stable amplifier.

Nyquist Criterion

Nyquist showed that the stability of a feedback amplifier depends on whether the response curve encircles the "critical point". The critical point is where the gain around the loop is exactly one. Note that the point was drawn as (1,0) in Nyquist's paper but is now usually drawn as (-1,0).
That loop gain exactly equal to one will produce an oscillation makes intuitive sense, as is the observation that the closer the response curve approaches the critical point then the less the stability. It is essential not to be too close to instability, and not just for the obvious reason of component tolerances. These tolerances obviously do set a limit but a practical circuit will have its behaviour controlled by well-defined resistors and capacitors rather than loosely defined semiconductor parameters, so actual tolerances can be small.
In practice the instability limit is set more by inevitable undesirable behaviour near instability.
This behaviour is most obvious in the time domain response of the circuit and so circuits are usually checked for instability in that domain, with an oscilloscope for a physical circuit or the equivalent in a simulation. The most obvious symptom is overshoot in the step response of the circuit that becomes worse as the system approaches instability.

There are rules of thumb about the level of overshoot that is acceptable as an indicator of incipient instability but these rules depend on assumptions that are often skipped over, and sometimes inapplicable. The rules and subtleties in the interpretation of time domain behaviour and its relation to stability will be examined more closely later, after some more foundations are in place.
The closeness to instability is commonly specified with two parameters, the phase when the loop gain is 1, abbreviated as PM, and the gain when the loop phase is 180, abbreviated as GM.
These parameters are more oriented to other plots, on a Nyquist plot the closeness of the response to the critical point and to instability can be seen directly. In practice the two parameters are usually sufficient.
Simpler tests of stability have been developed but the so-called "Nyquist criterion" remains the definitive test because it successfully handles time delays and other complications.

Conditional Stability

One complexity that can not be skipped is "conditional stability" because it is a concept that is important to understand for advanced amplifier compensation.
Conditional and unconditional stability are terms that are often misunderstood, especially in the discussion of conventional dominant pole amplifiers where they are not very relevant concepts.
Consequently the term "unconditional stability" has been free to be misused. Even the misuse has not been consistent, sometimes used to describe stability with various loads but usually for what is accurately termed "unity gain stability". Unity gain refers to the closed loop gain of an amplifier obtained with maximum negative feedback so it is no surprise that this is a worst case for stability (figure 2). In contrast, conditional stability refers to the case that an amplifier can become unstable if the negative feedback is reduced. Nyquist was the first to describe the possibility of this special case and it is useful to reserve a distinct term for the condition.

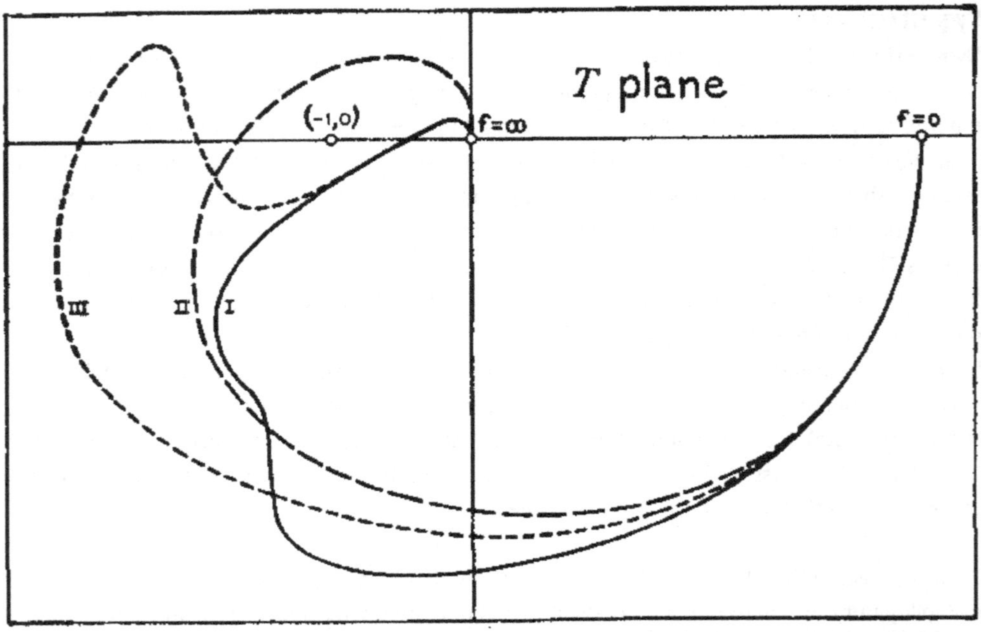

Figure 2 The Nyquist plot for a conditionally stable amplifier, III compared to a stable, I and unstable, II amplifier; from Bode.

Conditional stability was a major problem for the vacuum tube amplifiers of Nyquist's era because the gain would slowly increase as the valves warmed up. An instability at low gain would have time to establish an oscillation and the amplifier often became stuck in that state. For solid state amplifiers it is much less of a problem because nominal gain can be established almost instantly. Sometimes measures are taken to ensure that amplifier sections are powered up in a certain sequence to minimise problems.

Reduced gain and consequent risk of oscillation from conditional stability can also be a problem when an amplifier clips. There exist mathematical tools to handle this sort of instability but in practice such problems are better tackled with simulations. In particular the .TRAN analysis is used and will be covered in more detail later. The acceptable limit of conditional stability is a practical rather than theoretical decision, based on the particular application. My personal recommendation is to keep an amplifier unconditionally stable but use the freedom to come close to conditional stability.

Bode Plot and Minimum Phase

H. Bode [4] was the last of the trinity of Bell Labs researchers and completed the theoretical foundations of feedback theory with a work that made major steps in several distinct areas.

One of the foundations was to clarify the relation between amplitude response and phase. The fact that there is an unavoidable relation is not intuitively obvious, it seems it should be possible to roll-off the amplitude response and not alter the phase. However experiments and earlier study had shown that a low pass roll-off, for instance, always caused a phase delay to some extent.

This is most easily shown on the well-known Bode plot, which displays amplitude and phase as separate lines on the Y-axis with phase plotted linearly and amplitude plotted logarithmically, or in other words for math-phobes, in decibels, see figure 3. The X-axis is frequency plotted logarithmically.

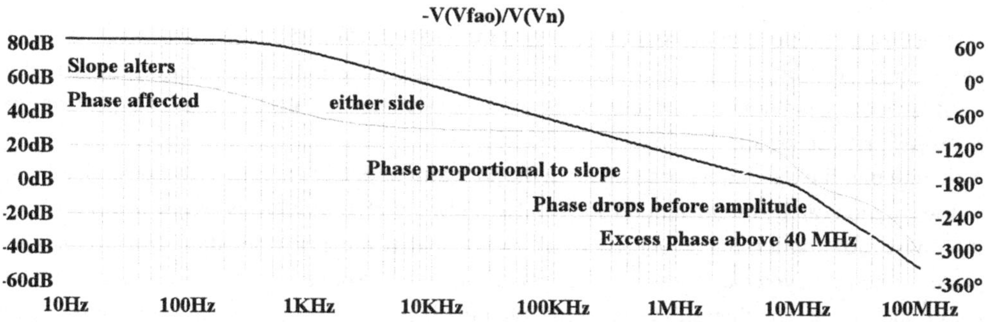

Figure 3 Bode Plot of a typical amplifier illustrates the relation between amplitude response and phase.

Bode realised that there was always a minimum amount of phase variation caused by a certain amount of amplitude roll-off and introduced the concept of a "minimum phase" system. For such systems the phase response can be derived from the amplitude response. It is however possible to have more than the minimum phase, this is the strict interpretation of "excess phase". The earlier work had been confused because of a failure to separate the two components. The distinction between the minimum phase and the excess phase is quite important in advanced compensation schemes so be aware that even some modern works use the term loosely

Bode calculated what mathematical operation is needed to derive the minimum phase from the amplitude. He seems never to have used the term but the operation is known in mathematics as a Hilbert transform and has similarities to the more familiar Fourier transform.

The Hilbert transform, despite its impressive name, is actually pretty easy to become comfortable with because it is a sort of derivative and acts like one. The phase at a specific frequency is determined by the slope of the amplitude curve, not exactly at the frequency as for a simple derivative, but mostly at and near the frequency. The influence of the slope of other frequencies tails off the further away they are and is small at about a factor of 10 from the specified frequency.

The effect is far simpler to see on a Bode plot than to describe, so just look at example plots until you can estimate the phase from the amplitude curve by eye. Figure 3 shows a Bode plot of the same typical amplifier used in the first Nyquist plot.

Minimum phase systems have several very important properties that sets them apart from non-minimum phase systems. Firstly, their effects are reversible. If a minimum phase system has an uneven amplitude response then the phase plot will have bumps and dips too. If that system is now cascaded with another minimum phase system such that the amplitude response is equalised then the phase plot will automatically be flattened too, the phase transform is reversible. This is not the case for excess phase; the delay can't be reversed, at least not without a time machine. The term used for this is "causal" - all effects have causes only in the past. The fact that a stable, causal minimum phase system has a stable, causal inverse is sometimes used as the definition of minimum phase. This reversibility has practical implications and also makes the mathematics much simpler.

Another useful property of minimum phase systems is that the time domain response can be completely predicted from the amplitude response. So two amplifiers with the same "frequency response" must have the same time domain behaviour too, provided they stay linear and are both minimum phase, as they usually are.

Also note that the minimum phase variation produced for a specific amplitude response is unique, so the function can be turned around and used to derive the shape of the amplitude response from the phase plot, though the actual amplitude response may differ by a constant. Since minimum phase is important you may have wondered about whether "maximum phase" exists. It is possible to define such a term and then subdivide non minimum phase into mixed phase and maximum phase systems but the distinction is academic for most purposes and the terms are practically never used in the audio literature.

Audio amplifiers are usually close to minimum phase but excess phase is still of concern because it sets the ultimate limits. Excess phase comes from two main causes:

1 Feed-forward in the system can create what are called RHP (right-hand-plane) zeros. In a circuit this would typically be feed-forward through a capacitance, either the internal capacitance of a semiconductor but more noticeably in compensation capacitors and especially the Miller capacitor. RHP zeros are essentially never desirable. Most have a barely detectable effect but it is possible for a series of circuit choices, each individually unexceptional, to move an RHP zero down in frequency until it makes a difference. Such RHP zeros should be eliminated with simple modifications as discussed later.

2 Time delay is inherently irreversible and hence causes excess phase. This is the kernel of sense at the heart of the simplistic myth that feedback "is always late" and can only try to correct problems when it is too late. It is true that time delay sets an ultimate limit on how much feedback can be applied.

Multiple hi-frequency poles produce an effect that resembles excess phase. An amplifier will inevitably have hi-frequency poles that are difficult to quantify and they are often treated as excess phase. Excess phase in individual transistors is usually not well quantified or understood either but probably results from a combination of the above effects and is dealt with similarly.

Bode Plot and Stability

Much of Bode's work applies equally well to multi-loop amplifiers but note that the stability calculations are aimed towards an amplifier with only one feedback loop. The requirement for only one feedback loop is not as restrictive as first appears, there can be many alternate feedback paths in the system provided it is possible to find one node that they all traverse. At that node all the feedback loops are effectively in parallel and can be treated as one loop. This condition is true for many audio amplifier compensation schemes as illustrated in figure 4.

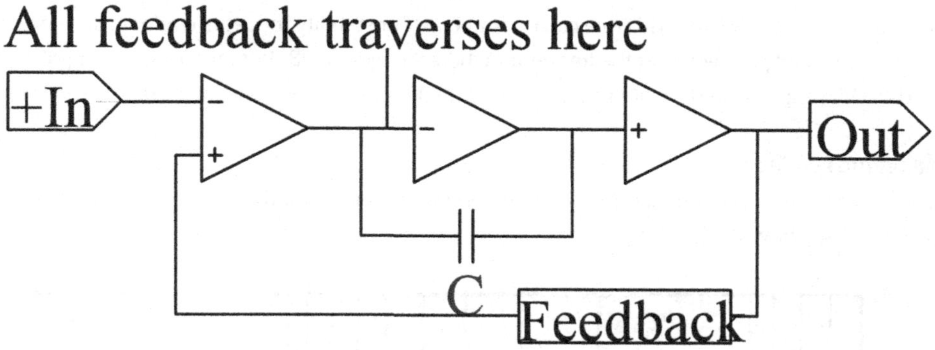

Figure 4 Block view of a typical Miller compensated amplifier with the correct node to check stability.

A Miller compensated amp appears to have two loops but they can be analysed as one because there is indeed a node they both traverse. This node is the correct place to determine the stability of the amplifier, what is remarkable is that it is not where a Miller amplifier is usually checked. The conventional test is usually adequate for a Miller amplifier but is actually incomplete. The question of where to check the stability will be further answered as we examine other compensation schemes.
The two loops of the Miller amplifier provide a clear illustration of the difference between loop gain and Bode's Return Ratio. Each loop could be checked individually for loop gain but the return ratio provides the best information about the stability of the amplifier as a unit. The two loop gains are effectively parallel elements for the return ratio (figure 5). The Bode plot provides a different way to look at stability from the Nyquist plot previously shown.

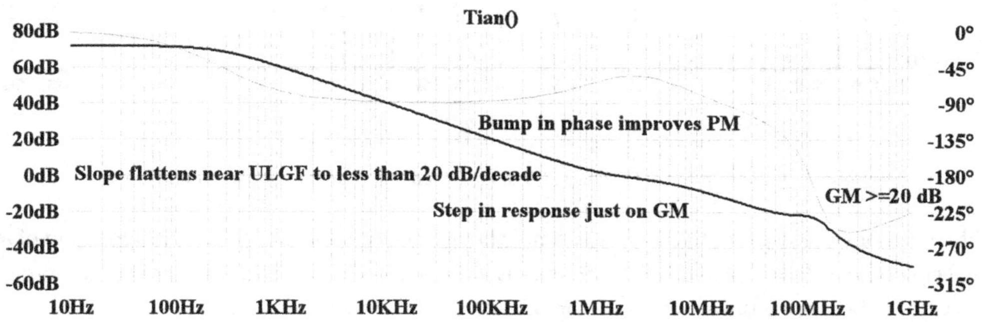

Figure 5 Bode plot of return ratio to show ULGF, slope and PM, GM when advanced techniques are used for conservative stability. Note that 0 dB lines up with 180°. This makes the plot easier to check.

The relation between amplitude and phase means that for a minimum phase system only the amplitude response needs to be considered and this simplifies the determination of stability.
Subject to certain conditions then the rule for stability becomes easy to state and, even better, is very easy to see on the plot. This rule can be stated as "For stability the return ratio must slope down at less than 40 dB per decade at the frequency where it crosses 0 dB". But for advanced amplifiers this tends to be simplistic and we are better to examine the phase near ULGF for extra information.

Maximum Feedback

Finally, with all the mathematical preparation done, Bode calculated the Maximum Obtainable Feedback curve as shown in figure 6.

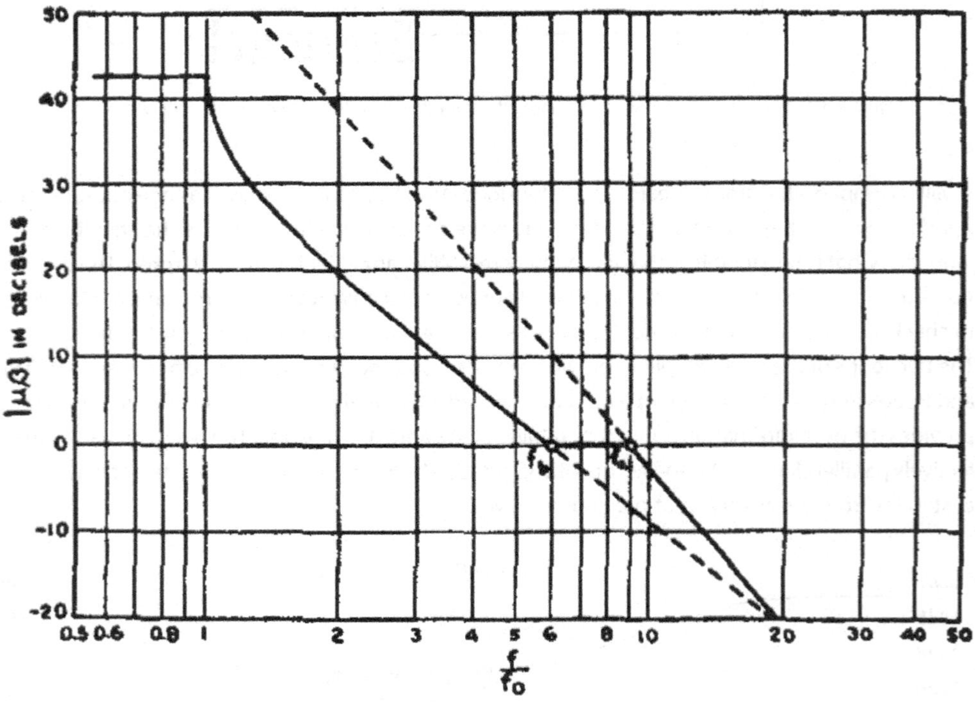

Figure 6 First illustration of Maximum Obtainable Feedback from Bode's foundational book.

The most obvious aspect of this curve is simply how different it looks to the actual loop gain of a Miller compensated amplifier. Every difference is an area where a conventional amplifier is sub-optimal, or to be more positive, where we can do better with an advanced amplifier.

Let us examine the curve in figure 6 from the highest frequency downwards.
The first point is that the position of the whole curve is set by the speed of the open loop system. This is not new; it is the truth at the heart of some audio myths. It is the reason that experts worry about

the performance of amplifiers up to frequencies that seem ridiculously high. But notice that the speed that matters is the maximum frequency where the open loop system still has gain. It is definitely *not* where the open loop system starts to roll off, as often claimed. The desirability of speed has clear implications at the circuit implementation level that will be detailed later.

The next feature to notice is the so-called "Bode step". This adds some phase lead at the ULGF to meet the stability criteria. The size of the step depends on the slope of the asymptote, the steeper the gain falls off then the more extended the step needs to be. The size of the step also depends on the desired slope of the loop gain. The step can be calculated from the formula in Bode but is typically done by simulation because the Maximum Obtainable Feedback curve is a type of mathematical function that can only be approximated in reality.
The Bode step ends at a zero and the loop gain starts to slope up at a rate dependent on the required phase, usually between 20 and 40 dB per decade. If the slope is 20 dB per decade then the phase in this area will be 90°. A slope more than 40 dB per decade will lead to conditional stability and this may be acceptable. Bode does not recommend any particular value. The use of 30 dB per decade in Cordell [7] is a somewhat oversimplified version of his ideas, but typical of practical amplifiers.

The final feature is the "Bode fillet" that allows the maximum loop gain to extend about an extra octave. The theoretical solution for the shape is impossible to implement in practice but some compensation schemes do have a peak here. This is usually considered a worry but Bode's result makes it look like a possible bonus.

Now, at last, we can start to look at some advanced compensation schemes and use the theory to understand them. The easiest scheme to start with is Two Pole Compensation (TPC) because it can be analysed as a one-loop feedback system, so the theory is directly applicable. Actually the system would be better called Two Pole and One Zero Compensation. The extra zero is important because it permits a closer approximation to Maximum Obtainable Feedback.

In fact the entire improvement of TPC can be explained simply by that closer approximation alone. That also provides a clue to where further improvements in TPC-like schemes can be made. An obvious possibility is to add a Bode step. What is normally called a phase lead capacitor placed across the feedback resistor can be considered an approximation of a Bode step. To view it from this perspective makes the value easy to calculate with the published Maximum Obtainable Feedback formula before it is fine tuned in the simulator. Caveats on this tactic will be explained in the circuit level discussion.
A Bode fillet is also a possibility. The analysis of TPC by Dymond [5] does not accurately predict the behaviour near this frequency but the study by Feucht [6] is more detailed, and of course more complicated.

Next we want to compare TPC with other compensation schemes like Transitional Miller Compensation, but this has been a contentious subject. To ensure a fair comparison requires some explanation in exactly how and where to measure the true return ratio. This is an important skill in any case, and so as well as further explanation of compensation schemes, the next instalment will cover:

- Linear stability tests with simple V source, Middlebrook 1975, Rosenstark, Tian, Middlebrook 2006 "Final Solution";
- Probe placement and broken loops, the Hurst balun technique;
- Implementations in LTspice and a little about Spectre, ICAP and other simulators.

References

[1] Putzeys, Bruno."The F-word" Linear Audio Vol. 1

[2] Black, H.S. Bell Laboratories Record. 1934

[3] Nyquist, H Bell Systems Technical Journal. 1932

[4] Bode, H.W. Network Analysis and Feedback Amplifier Design. 1945

[5] Dymond, H. "Analysis of Two-Pole Compensation in Linear Audio Amplifiers" 129th AES Convention

[6] Feucht, D. Two-Pole Compensation Parts1 and 2; www.analogzone.com

[7] Cordell, R.R. Designing Audio Power Amplifiers. 2011

[8] Self, D. Audio Power Amplifier Design Handbook. 5th Ed. 2009

Recent investigations and developments related to feedback and feed-forward error correction - Part 1

Giovanni Stochino

Introduction

In this contribution I would like to present and discuss the results of my recent investigations and findings on some popular feedback and feed-forward error-distortion correction schemes, as well as some practical developments which have been incorporated in my latest low distortion audio power amplifier designs.

Error correction (EC) is something different from the distortion reduction attained with standard overall feedback application (FDR – Feedback Distortion Reduction), and is the generalization of the *distortion correction* concept, formally introduced by M. J. Hawksford in 1981 **[1]** and further developed in 1983 **[2]**. However, its historical background dates back to the pioneer works by H. B. Black in 1928 **[3] [4]**, and later in 1933 by Baggally **[5]** and in 1955 by Macdonald **[6]**.

In Part I of this contribution, I will first review a few feedback error correction solutions which I have found effective and practical to apply to audio power amplifiers:

- the active-error feedback scheme (AEF) which was first used for error-correction in amplifiers by J. R. Macdonald in 1954;
- the Hawksford feedback error correction scheme (HFEC), which is very popular among audio design engineers. There are indeed some interesting implementations already available, e.g. by R. Cordell **[7]** , by J. Didden **[8]**, by B. Halcro Candy **[9]**, and by O. Popa and E. Stuart (http://home.tiscali.nl/audio/index.html).

The distortion correction technique introduced by E. Cherry (**[10]** and **[11]**) and based on the use of multiple nested feedback loops (NDFL), is outside the scope of this discussion and will not be considered. I will examine these EC schemes not only from a system perspective, but also from an engineering and implementation standpoint, with the support of XSPICE (Multisim) simulation graphs and experimental data.

The points that will be more specifically taken into account are the following:

- time and frequency response (or , in short, time-frequency response) and practical stability criteria;

- error correction potential;
- weaknesses and strengths, as well as some recommendations and implementation aspects.

I will eventually introduce and discuss a new EC scheme that has been named *Balanced Error Feedback*, in short BEF (Italy and US patent pending) **[12]**. The BEF scheme will be compared with the concepts previously examined, and its advantages in terms of time-frequency response, stability and error correction performance will be highlighted. We will see in more detail a few representative experimental implementations incorporating a BEF error-correction loop: a MOSFET power output stage (gain≅1) and a simple but complete 200W MOSFET power amplifier (gain≅30). I will finally address BEF's distinctive properties of easy iteration in order to reduce distortion to negligible levels.

In Part II, in a future Linear Audio article, I will present and discuss the recent developments of the true feed-forward error correction (FFEC) implementation (in full accordance with Black's original scheme) which was first published in 1994 **[13]**, where a small ferrite wideband transformer, incorporating a main signal flux cancellation technique, was used in the output summer of a feed-forward 100W audio power amplifier. These developments have been employed in the *superLinear* 200W audio power amplifier demo-system (Pre-Amplifier and Power Amplifier), to achieve a total harmonic distortion lower than 0.0005% and an intermodulation distortion of less than 0.0001%.

Finally, we will have a closer look at a new development for both BEF and FFEC built in a synergetic way around a hybrid scheme to potentially achieve harmonic and intermodulation distortion levels stably lower than 1ppm (0.0001%) at all audio frequencies and power levels in audio power amplifiers.

Error Correction

In the following discussion I will use *"Error Correction"* (EC) to indicate a process that specifically addresses the output error voltage $\mathbf{V_e}$ of the *target* amplifier (the amplifier to be corrected), intended as the difference $\mathbf{V_e=V_O-V_{OA}}$ between the wanted value V_O and the actual value V_{OA} of the output voltage. $\mathbf{V_e}$ is processed by the EC circuit in a logical sequence of actions including:

1 accurate error voltage sensing and extraction;
2 production of a correction voltage $\mathbf{V_C}$, which is as accurate as possible a copy of the error voltage $\mathbf{V_e}$;
3 addition of the correction voltage $\mathbf{V_C}$, with the right polarity, to either the input voltage or the output voltage of the target amplifier, in order that the overall output error is virtually cancelled;
4 none or very limited impact of the correction process on the transfer function of the target amplifier.

Action 4 is an additional requirement, not usually included in the original EC theory. It implies that the EC process should *more specifically* address the **non-linear error/distortion** of the target amplifier (the artifacts added by the target amplifier but not present in the input signal, as harmonics and intermodulation products) **and possibly its DC and low frequency error (offset and dc gain), drift and noise**, but **without altering its time-frequency response, as well as its large-signal dynamic performances, e.g. slew rate and power bandwidth.** Any impact of EC on the **linear-error** of the target amplifier (the magnitude and phase of its input-output transfer function) will in general be considered an *unintended and undesired side effect* of EC. Meeting this fourth requirement entails that the input-signal dependent component at the input of the target amplifier remains unchanged, at all frequencies and voltage levels, when the correction process is activated. The important end result of this requirement, as we will see later, is to improve the performance of the error correction process and to extend the range of effective application of the EC scheme.

The model of the target amplifier. In this context , the target of the EC process is a generic audio power amplifier (class A or AB, with bipolar or MOSFET power devices), named $\mathbf{A_{1T}}$. It can be either the whole amplifier or part of it, like for instance the output stage. It is assumed to have a positive voltage gain A_1 (A_{10} at DC), nominally equal or greater than one, and good stability margins. It is also assumed to have reasonably high input impedance and low output impedance in its entire almost-linear operating range. Its simplified model I will use in this discussion is shown in **Fig.1**.

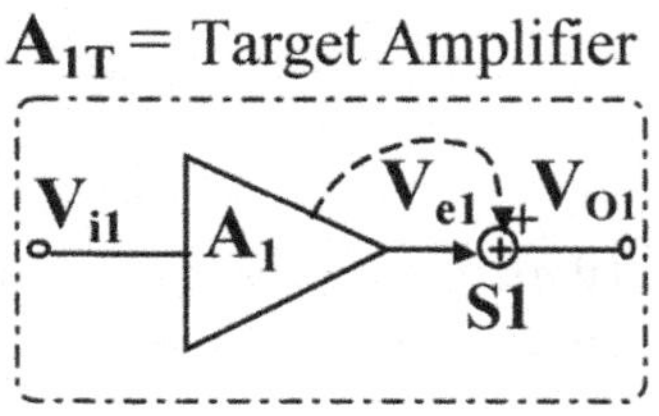

Figure 1

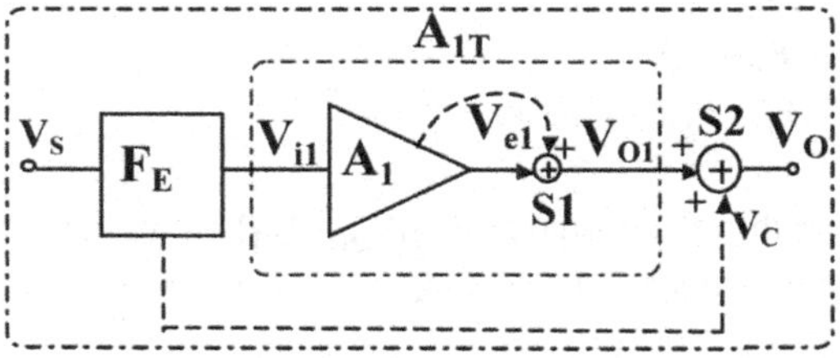

Figure 2

Figure 1 *The model of the voltage amplifier* $\mathbf{A_{1T}}$, *target of the error correction process*
Figure 2 *Simplified equivalent block diagram of an error-corrected amplifier.*

In the simple model of Fig.1, the real amplifier $\mathbf{A_{1T}}$, which produces the nonlinear error $\mathbf{V_{e1}}$ during its operation, is modeled by an ideal amplifier $\mathbf{A_1}$, free from nonlinear-error but having the same transfer function of A_{1T} and voltage gain A_1, and by the non-linear error voltage $\mathbf{V_{e1}}$ (distortion, offset voltage, drift and noise voltage, etc.), which is added to the output of $\mathbf{A_1}$ by the voltage summer **S1**. The output voltage therefore will be equal to

$$\mathbf{V}_{O1} = A_1 V_{i1} + V_{e1} \tag{1}$$

The application of error correction to our target amplifier $\mathbf{A_{1T}}$ has in general two main consequences:

- its transfer function will be modified, to a small or large extent depending on the ***dynamics*** of the correction process (linear error);

- the non-linear error will be reduced, to a small or large extent depending on the ***effectiveness*** of the correction process.

These effects of the error correction process can be conceptually modeled as depicted in **Fig.2**, where the F_{EC} block (which I call the *equivalent EC transfer function*) incorporates the changes of the overall transfer function of A_1 attributable to the error correction process, and V_C is the *equivalent* correction voltage (which also contains the error V_{ec} produced by the same EC process, usually much smaller than V_{e1}) that is added, via the voltage summer **S2**, to the output voltage of the target amplifier, with the appropriate phase, in order to cancel the error V_{e1}. Therefore the output voltage V_O of the overall corrected amplifier can be written as

$$V_O = F_{EC}A_1V_S + V_{e1} + V_C = F_{EC}A_1V_S + V_{eo} \qquad (2)$$

where $F_{EC}A_1$ is the overall transfer function of the corrected amplifier and $V_{eo} = V_C + V_{e1}$ is the residual error of the correction process. So, in accordance with my previous EC definition, the ideal situations occurs when

$$F_{EC} = F_{ECO} = 1 \qquad (3)$$

$$V_{eo} = V_C + V_{e1} = V_{ec} + V_{e1}/\rho \rightarrow 0 \qquad (4)$$

where the relationship $\rho = V_{e1}/V_{eo}(V_{e1})$ between the error V_{e1} of the target amplifier and its residual component at the output, $V_{eo}(V_{e1})$, is the ***error cancellation ratio*** (the higher, the better!).

EC classes. In the context of this contribution, the error correction (EC) techniques may be conveniently grouped into three classes:

I	**FBEC**	feed-back error correction;
II	**FFEC**	feed-forward error correction;
III	**FB-FFEC**	mixed feed-back and feed-forward error correction.

FBEC. In FBEC schemes, as shown in **Fig.3**, the error correction circuitry (**ECC**) senses the source voltage V_S [option (I)] which stimulates the full FBEC system, or the input voltage V_{i1} of the target amplifier A_{1T} [option (II)], and the overall output voltage V_O, and processes this information in accordance to its internal circuit configuration to generate the correction voltage V_C which is fed back with the appropriate phase [downstream of V_S in option (I) or upstream of V_{i1} in option (II)] to the input of the target amplifier, via the voltage summer **S2**. A negative feedback loop (nFBL) is then inevitably established in both options [(I), (II)], with the well-known stability problems that it brings along. In option (II) a positive feedback loop (pFBL) is also created, which may aggravate system stability. FBEC schemes are discussed in this first part.

FFEC. In FFEC schemes instead, as shown in **Fig.4**, the error correction circuitry (**ECC**) senses the source voltage V_S (=V_{i1}) and the output voltage V_{O1} of the target amplifier A_{1T}. This information is then processed by ECC to generate the corrective voltage V_C which is fed forward and combined with the output of the target amplifier via the voltage summer **S2**. In this way no feedback loop is established (at least in theory and in good implementations as well!) and the stability of the corrected amplifier is preserved. A correction performance superior to FBEC is therefore expected.

FB-FFEC. The basic block diagram of the error correction scheme which mixes both FBEC and FFEC is shown in **Fig.5**. Here **ECC** senses the source voltage V_S and the output voltage V_O, as in FBEC, but the correction voltage V_C is not fed back to the input of A_1, as in FBEC, but is fed forward and added up to the output voltage V_{O1} of the target amplifier A_{1T} via the voltage summer **S2**, as in FFEC. In this way a feedback loop (nFBL) is established again, however, contrary to FBEC scheme, the target amplifier A_{1T} is left outside the nFBL loop, so that stability problems of the corrected amplifier can be less critical, compared with FBEC, and correction performance could in theory be improved. FFEC and FB-FFEC schemes will be discussed later in the second part of this article.

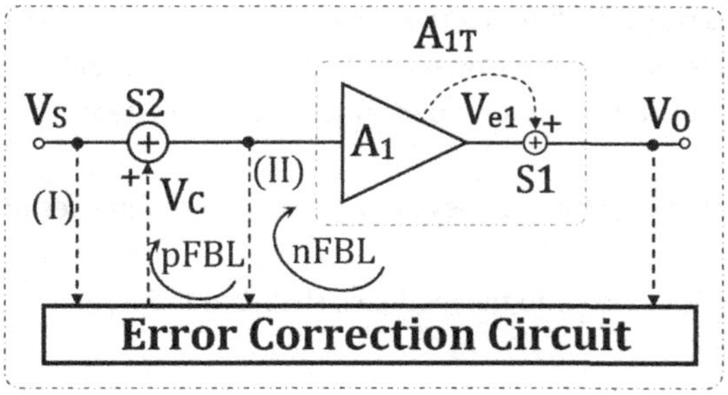

Figure 3 Simplified block diagram of a generic feedback error correction scheme (FBEC).

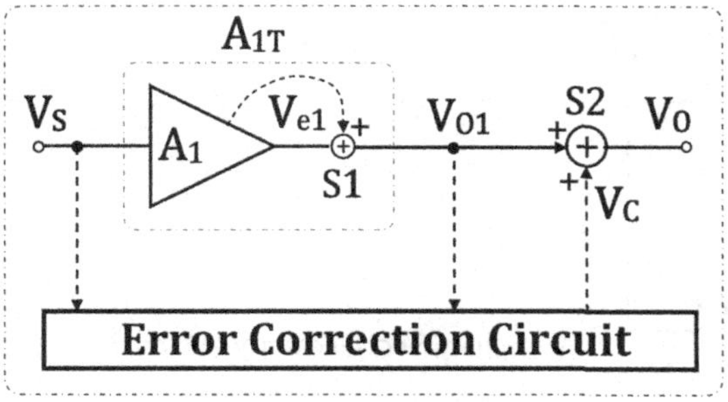

Figure 4 Simplified block diagram of a generic feed-forward error correction scheme (FFEC).

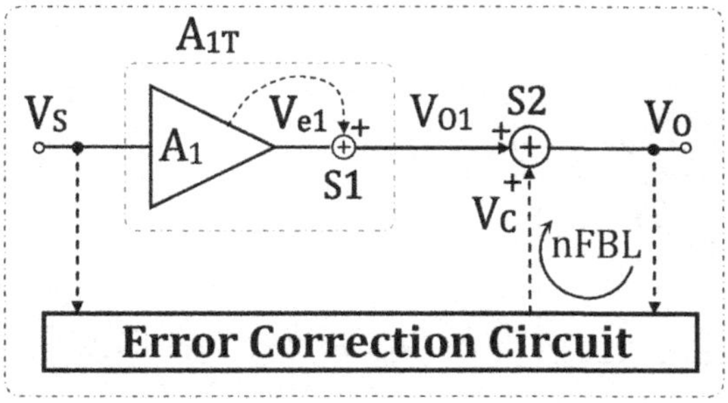

Figure 5 Simplified block diagram of a generic mixed FBEC and FFEC error correction scheme

Active Error Feedback (AEF) by Macdonald [FBEC option (I)]

The first EC scheme I will review, was named *Active Error Feedback* (AEF) by J. R. Macdonald in his important article published in 1955 **[6]**. The basic block diagram is depicted in **Fig.6.** The AEF EC circuit is composed of tree functional blocks: the extra-gain (and error) amplifier $\mathbf{A_2}$ (with $A_2 \to \infty$), the voltage summers **S3** and **S4.** $\mathbf{V_{e2}}$, and relevant summer **S2,** is the overall error associated with the added blocks and inevitably introduced into the game. In this case our target amplifier $\mathbf{A_{1T}}$ has a nominal unity gain, that is $A_1 \approx 1$ (e. g. the output stage of an audio power amplifier). An immediate comparison with the standard *feedback distortion reduction* scheme (FDR) depicted in **Fig.7** (which, among other things, is electrically and functionally equivalent to the nFBL loop of the AEF scheme), allows us to appreciate the distinctive feature of the AEF scheme (yet common to all FBEC systems) where the input signal $\mathbf{V_{s,}}$ via S4, always retains an active and direct part in the control of the amplifier to be corrected. The same does not occur in ordinary FDR systems, since the control of $\mathbf{V_s}$ on the target amplifier is always mediated by the amplifier A_2 (with all limitations of its transfer function and dynamics). This specific feature of FBEC schemes has very important consequences in the time-frequency response of the full corrected system, since at high frequency the amplifier $\mathbf{A_2}$, with its low gain (*low* for stability reasons) and high phase lag, is completely bypassed by the source of input signal $\mathbf{V_s}$, which resumes the full control on $\mathbf{A_{1T}}$ and on the dynamic performance of the system.

So, although AEF and FDR have exactly the same gain blocks, they show important differences in their configurations and operations:

- in FDR, A_2 is cascaded with A_{1T}. In AEF, A_2 is moved into the feedback chain around A_{1T};

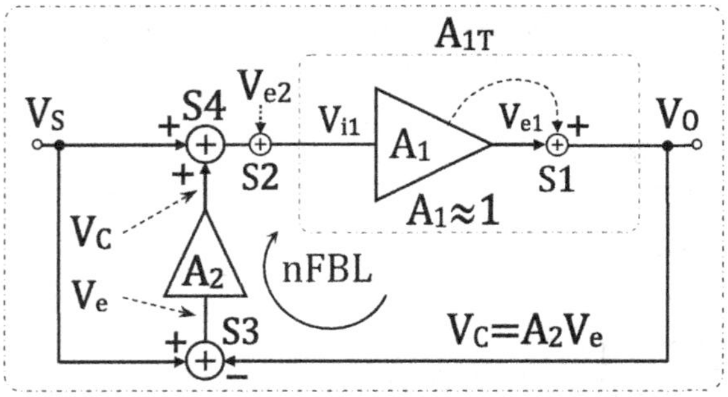

Figure 6 Block diagram of AEF scheme.

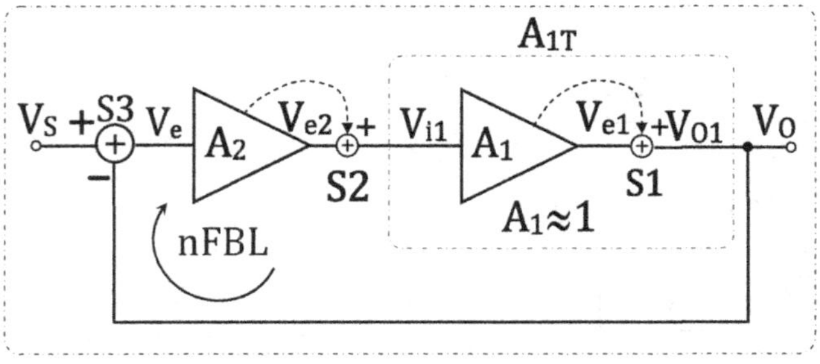

Figure 7 Block diagram of the standard FDR scheme.

- in FDR, A_{1T} is driven by A_2 output voltage only, while in AEF A_{1T} is driven by the sum of the system input voltage V_S and the output voltage of A_2;
- in both schemes the *total error*, $\mathbf{V_e=V_S-V_O}$ (with linear and non-linear components) is amplified by A_2, but only in AEF is added to the source voltage, as expected by FBEC, to drive the target amplifier.

However, there are also some similarities to highlight. In both schemes in fact:

- the error-reduction operation is based on negative feedback and relies on the accuracy and the high open loop gain of the correction loop in order to reach the goal of $\mathbf{V_e=0}$, and there is no need for calibration;
- the error correction loop gain is equal to A_1A_2. To see it, simply put $\mathbf{V_S=0}$, and follow the paths of V_{e1} and V_{e2} in the nFBL loop in both schemes. Therefore, stability problems of the EC loop are expectedly similar;
- block cascade $S3\times A_2$ can be implemented with a simple Op-Amp in both schemes.

I will now examine and concurrently compare AEF and FDR analytically, with the essential mathematical steps. With simple calculations I first determine the output voltage of the two schemes, which are

$$\mathbf{V_o}(FDR) = V_sA_1\frac{A_2}{1+A_1A_2} + V_{e1}\frac{1}{1+A_1A_2} + V_{e2}\frac{A_1}{1+A_1A_2} \tag{5a}$$

$$\mathbf{V_o}(AEF) = V_sA_1\frac{1+A_2}{1+A_1A_2} + V_{e1}\frac{1}{1+A_1A_2} + V_{e2}\frac{A_1}{1+A_1A_2} \tag{5b}$$

The overall transfer function $\mathbf{V_O/V_S}$ and relevant EC transfer functions $\mathbf{F_{EC}}$ are then given by

$$\frac{\mathbf{V_o}}{V_s}(FDR) = A_1\frac{A_2}{1+A_1A_2} = A_1F_{EC}(FDR) \tag{6a}$$

$$F_{EC}(FDR) = \frac{A_2}{1+A_1A_2} \tag{6b}$$

$$\frac{\mathbf{V_o}}{V_s}(AEF) = A_1\frac{1+A_2}{1+A_1A_2} = A_1F_{EC}(AEF) \tag{6c}$$

$$F_{EC}(AEF) = \frac{1+A_2}{1+A_1A_2} \tag{6d}$$

and the residual output error $\mathbf{V_{eo}}$, which is exactly the same for both schemes, is equal to

$$\mathbf{V_{eo}} = V_{e1}/\rho_1 + V_{e2}/\rho_2 \tag{7}$$

where $\rho_1=(1+\mathbf{A_1A_2})$ and $\rho_2=(1+\mathbf{A_1A_2})/\mathbf{A_1}$ are the **error cancellation ratios** for $\mathbf{V_{e1}}$ and $\mathbf{V_{e2}}$, respectively.

Since $A_1 \cong 1$, for $A_2 >> 1$ we also have that

$$\rho_1 \cong \rho_2 = \rho = 1 + A_1 A_2 \cong A_2 \tag{8}$$

which states that the error cancellation ratio for both AEF and FDR is virtually equal to the gain of the added amplifier A_2 (of course when system stability is ensured). Equations (5a) and (5b) tell us that the two schemes also share the same fundamental stability condition to be met by all feedback systems, which states that the denominator of the transfer function has to be always different from zero, that is

$$1 + A_1 A_2 \neq 0 \tag{9}$$

If we now take a closer look at equations (6a) and (6c), we see that the only difference between the two transfer functions , is represented by the term 1 in the numerator of F_{EC}(AEF) which is missing in F_{EC}(FDR). This apparently minor difference has an important effect on the overall response of the two EC systems. Considering, in fact, that at low frequency $A_2 \to \infty$ and at very high frequency A_1 & $A_2 \to 0$, both FDR and AEF have the same low frequency behavior, that is

$$\lim_{f \to 0} \boldsymbol{F_{EC}}(FDR) = \lim_{f \to 0} \boldsymbol{F_{EC}}(AEF) = \mathbf{1} \tag{10}$$

in line with the target $F_{EC}=1$. At the highest frequency the behavior of the two schemes is substantially different, as expressed by the following limits

$$\lim_{f \to \infty} \boldsymbol{F_{EC}}(FDR) = \boldsymbol{A_2} \tag{11a}$$

$$\lim_{f \to \infty} \boldsymbol{F_{EC}}(AEF) = \mathbf{1} \tag{11b}$$

which imply that F_{EC}(FDR) diverges from the ideal value at the same rate as A_2, while F_{EC} (AEF) tends, as we will see more clearly in the following, to remain close to 1 in a wider frequency range. If we refer to the 4th requirement of the EC definition, AEF has then to be considered a better error-correction scheme than FDR.

Time-frequency response. In order to get a deeper understanding of the time-frequency response and error-correction performance achievable by AEF and FDR schemes in Fig.6 and Fig.7, as well as of their relevant design and implementation implications, it is now necessary to take into account more specifically the transfer function of both the target amplifier A_{1T} and the extra-gain, or error amplifier A_2.

Let us therefore assume that both A_{1T} and A_2 have a *first order transfer function*, and more specifically

$A1 = 1/(1+s/p_1)$ with **$p_1 = 2\pi f_{C1}$** (12)

$A_2 = A_{20}/(1+s/p_2)$ with **$p_2 = 2\pi f_{C2}$** (13)

where $p_2 << p_1$, and $A_{20} \rightarrow \infty$ is the dc gain of A_2, which therefore, in association with S3, can be more clearly seen as an Op-Amp. Frequencies f_{C1} **and** f_{C2} are the corner frequencies (at -3dB) of A_1 and A_2 respectively.
Let's note that $A_{20}p_2 = p_{U2} = 2\pi f_{U2}$, **where** f_{U2} **is the unity gain frequency of** A_2.

We can now substitute eq. (12) and eq. (13) into eq. (6b) and eq. (6d), develop with standard calculation rules and achieve the equivalent EC transfer function F_{EC} for both FDR and AEF schemes, i.e.

$$F_{EC}(FDR) = \frac{A_2}{1 + A_1 A_2} = \frac{1 + s/p_{U2}}{1 + s/p_{U2} + s^2/p_1 p_{U2}} = \frac{1 + s/p_{U2}}{1 + (2\zeta/\omega_N)s + s^2/\omega_N{}^2} \tag{14a}$$

$$F_{EC}(AEF) = \frac{1 + A_2}{1 + A_1 A_2} = \frac{(1 + s/p_1)(1 + s/p_{U2})}{1 + s/p_{U2} + s^2/(p_1 p_{U2})} = \frac{1 + s(p_1 + p_{U2})/\omega_N{}^2 + s^2/\omega_N{}^2}{1 + s(2\zeta/\omega_N) + s^2/\omega_N{}^2} \tag{14b}$$

which are second order transfer functions, where $\omega_N = 2\pi f_N = \sqrt{(p_1 p_{U2})}$ is the *natural (angular) frequency* of the systems, and ζ is the *damping ratio*, which is equal to

$$\zeta = 0.5\sqrt{p_1/p_{U2}} = 0.5\sqrt{f_{C1}/f_{U2}} > 0 \tag{15}$$

It can be now highlighted that, as described above, both eq.(14a) and eq.(14b), are different from the ideal value, $F_{EC}=1$ (0dB), but F_{EC}(AEF) moves away from unity only in proximity of $f_N=\omega_N/2\pi$ where it reaches its peak given by

$$F_{EC}(AEF)(\omega_N) = (p_1 + p_{U2})/p_1 \tag{16}$$

which yields a peak of 6 dB for $f_{U2}=f_{C1}$ (i.e. $\zeta=\zeta_O=0.5$), of 12 dB for $f_{U2}=3f_{C1}$ and of only 2.3 dB for $f_{U2}=0.3f_{C1}$.
More precisely, for $f_{U2} \leq 0.3 f_{C1}$ AEF scheme is quite close to fully meet the 4th EC requirement, since the maximum deviation of F_{EC}(AEF) from its ideal value is only 2.3 dB in magnitude and 7.5 degrees in phase.

Stability issue. Eq. (15) is of paramount importance for the entire FBEC class of feedback error correction systems. Let's see why. As is well known, a second order system is stable if $\zeta > \mathbf{0}$ and in particular has acceptable time (none or low overshoot, low settling time) and frequency (none or low peaking) response, when, say, $\mathbf{0.3 < \zeta < 1}$.
Therefore **eq.(15)** states **two important facts:**

- when A_1 and A_2 are first order systems (but we must be aware that **this is never the real case!**), both FDR and AEF schemes are always stable (although **critically stable** with $\zeta < \mathbf{0.1}$);
- the actual damping ratio only depends on the ratio between the corner frequency f_{C1} of $\mathbf{A_1}$ and the unity gain frequency f_{U2} of $\mathbf{A_2}$ **(very simple and important information for designers).**

A reasonable approach, from a more practical engineering perspective, is that in designs incorporating one or more EC feedback loops, in order to limit the risk of instability it is always set $\zeta \geq \zeta_O$ for each loop, where ζ_O is a **reference** and limit value cautiously assumed to be ζ_O=**0.5.**

This precautious attitude is mostly driven by two factors:

- the foreseeable existence of **other high frequency poles** in the actual transfer function of both A_1 and A_2 (multistage) amplifiers in Fig.6 and Fig.7, which certainly end up reducing the stability margins;
- **the corner frequency f_{C1}** of the target amplifier in general is not constant and **depends on the position of the operating point of the amplifier.** In a power MOSFET output stage for instance, f_{C1} can change from about 20 MHz in the idle state (V_O=0, I_O=0, and $f_{C1}=f_{C1O}$) to about 3-5 MHz when V_O moves close to the supply voltages(where $f_{C1}=f_{C1W}<f_{C1O}$), due to the important increase of MOSFET internal capacitances with low drain-source voltage. A similar phenomenon occurs with output stages employing bipolar power transistor, whose transition frequency f_T, and then f_{C1}, strongly depends on the collector current. This implies that, for a fixed value of f_{U2} of A_2, **the damping ratio ζ moves up and down** accordingly, and stability may become critical if ζ **goes substantially below the safe limit of 0.3.**

Therefore I want to underline the **important and general relationship that links the unity gain frequency f_{U2}** of the additional amplifier **A_2 to the corner frequency f_{C1} of the target amplifier A_{1T} in a feedback error-correction loop**, when adequate stability margins, and hence good closed-loop time-frequency response, are required in all operating conditions. This simple relation is given by

$$f_{U2} \leq f_{C1}/(4\zeta_0^{\,2}) \qquad (17)$$

where ζ_O =**0.5**. With $\zeta \geq$ **0.5**, and hence $f_{U2} \leq f_{C1}$, **both** AEF and FDR distortion correction loops of Fig.6 and Fig.7, have adequate stability margins. In particular, when A_{1T} and A_2 have a *first order transfer function,* the phase margin is always higher than 52 degrees and the gain margin is always higher than 40dB.

Error-correction potential. In this context we can define the *error cancellation potential,* or more in general ***error correction potential*** ρ_P , as the amount of correction of the error/distortion of the target amplifier, V_{e1}, that can be achieved by FDR and AEF systems with adequate phase margins, that is when $f_{U2}=f_{C1}$. For this calculation we will assume V_{e2}= 0. If we substitute eq. (13) into eq. (8) and consider that for $f > f_{C2}$, $|A_2| \cong f_{U2}/f$, then the magnitude of ρ_P can be written as

$$|\rho_p| \cong |A_2| \cong f_{U2}/f = f_{C1}/f \leq A_{20} \qquad (18)$$

that is the maximum error reduction achievable at a given frequency *f*, **is equal to the ratio between the bandwidth of the target amplifier and the frequency *f*.** Therefore, for examples, if f_{C1}=10MHz, the third harmonic of a 20kHz input signal (i.e. at 60kHz) generated by A_{1T} can be potentially cut by 44dB in both schemes. If f_{C1}=1MHz instead, the reduction of the same 60 kHz harmonic cannot exceed 24dB. So, in a power MOSFET output stage, where, as mentioned above, f_{C1} can vary between 3MHz and 20MHz, the value of achievable distortion reduction of the third harmonic at 60kHz, is expectedly between 34 and 50 dB, depending on the actual value of f_{U2} set in the design phase.

XSPICE simulation. Let's have a look now at some enlightening XSPICE simulations graphs, assuming that our target amplifier A_1 has nominal unity gain and average corner frequency f_{C1}**=10MHz**, which is a typical value for a power N&P channel MOSFET output stage in voltage-follower (VF) configuration, and A_2 has a dc gain of A_{20}**=106dB**.

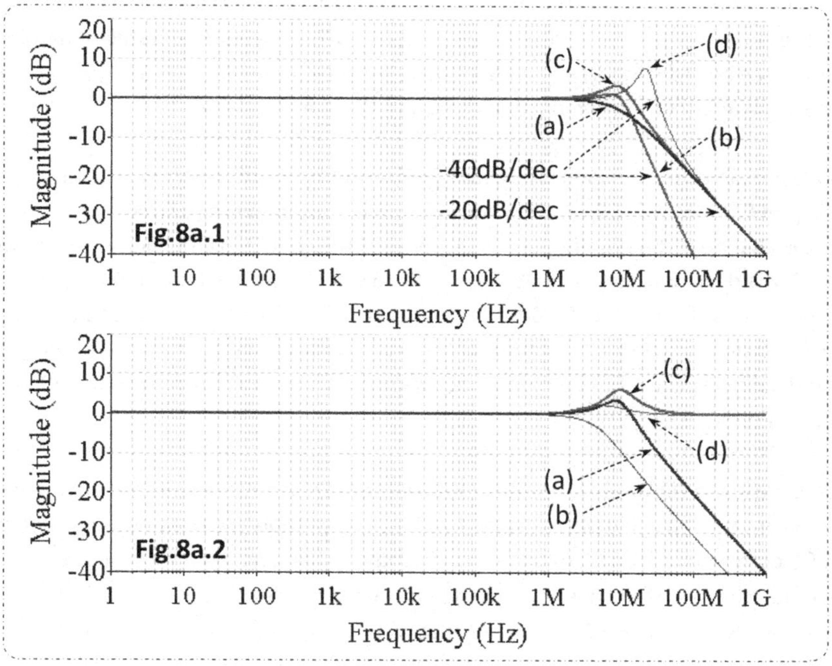

Fig.8a.1 FDR and AEF transfer functions (magnitude in dB)
- $f_{U2}=f_{C1}$**=10MHz (a,b,c):** (a) target amplifier (A_1); (b) FDR amplifier [F_{EC}(FDR)A_1]; (c) AEF amplifier (F_{EC}(AEF)A
- $f_{U2}=5f_{C1}$**=50MHz (d):**(d) AEF amplifier [F_{EC}(AEF)A_1]

Fig.8a.2 EC transfer function F_{EC} (magnitude in dB)
- (a) F_{EC}(FDR), $f_{U2}=f_{C1}$=10MHz; (b) F_{EC}(FDR), $f_{U2}=0.3f_{C1}$=3MHz
- (c) F_{EC}(AEF), $f_{U2}=f_{C1}$=10MHz ; (d) F_{EC}(AEF), $f_{U2}=0.3f_{C1}$=3MHz

Figure 8a Frequency response (magnitude in dB) of Fig.6 (AEF) and Fig.7 (FDR).

In **Fig.8a.1** we can observe the magnitude (dB) of the frequency response of the FDR and AEF error-corrected amplifiers compared with the original response of A_{1T} (alone) when $\boldsymbol{f_{U2}=f_{C1}}$ and $\zeta=\zeta_O$. We can see the expected bandwidth extension and the peaking of about 1.2dB for FDR and 3.4 dB for AEF, but what it is important to stress is the slope change of FDR system gain beyond about $1.5f_{C1}$, which compared to A_1 increases from -20dB/dec to a very critical -40dB/dec (see curve b). This slope increase is of extreme importance when we intend to **incorporate our FDR error correction circuit inside the entire audio amplifier** with its own overall feedback loop. Gain slope increase combined with increased phase lag, will in fact increase the risk of instability. In the AEF system instead, **and this is the main advantage of AEF scheme**, the risk of instability is lower, since the critical gain slope of -40dB/dec is reached only immediately after f_N and for $f_{U2}/f_{C1}>1$ (see curve d). For $f_{U2}/f_{C1}\leq 1$ instead, AEF gain slope is always lower than -40dB/dec and the AEF frequency response coincides with A_1 over most of the frequency band. This behavior may reduce considerably the risk of instability in amplifiers using AEF error-correction schemes properly. **When necessary, stability margins can be easily improved by reducing f_{U2}/f_{C1} ratio,** but we must be aware that this design solution has the side effect of lowering the error reduction capability of our AEF scheme.

In **Fig.8a.2** we can observe more closely, for $f_{U2}=f_{C1}$ and $f_{U2}=0.3f_{C1}$, the equivalent frequency response $\mathbf{F_{EC}}$ of FDR and AEF EC circuitry, which in the EC system model of Fig.2 holds the responsibility for the change in the time-frequency response of the corrected amplifier. In curves (a) and (b) of F_{EC}(FDR) the deviation from the 0dB line (the ideal value), starts at about $f=0.1f_{C1}$ and increases rapidly and constantly at a rate of -20dB/dec for frequencies above f_N. Curves (c) and (d) show that F_{EC}(AEF) starts deviating from the 0dB line about one decade before f_N as well, reaching peaks at f_N (2.3 dB and 6 dB, respectively, as discussed above) but, contrary to FDR, returns to 0dB in less than one decade after f_N.

This difference in the FDR and AEF frequency responses of Fig.8a.1, are reflected into their different time response depicted in **Fig.8b**. In **Fig.8b.1** with $f_{U2}=f_{C1}$=10MHz in both systems (therefore with optimal EC performance), AEF exhibits a significantly higher overshoot (30%) than FDR (16%). In order to reduce the overshoot it is necessary to reduce the f_{U2}/f_{C1} ratio, but the effects are different for the two systems. In **Fig.8b.2** we can see that with $fU2=0.3f_{C1}$ =3MHz only FDR has an optimal time-response without overshoot, although both AEF and FDR systems have similar settling performance. I have to stress at this point the big limit of the AEF scheme, which for space reasons cannot be discussed in depth here: overshoot is intertwined with settling time and can never be completely removed, even for f_{U2}/f_{C1} ratios much lower than 1.

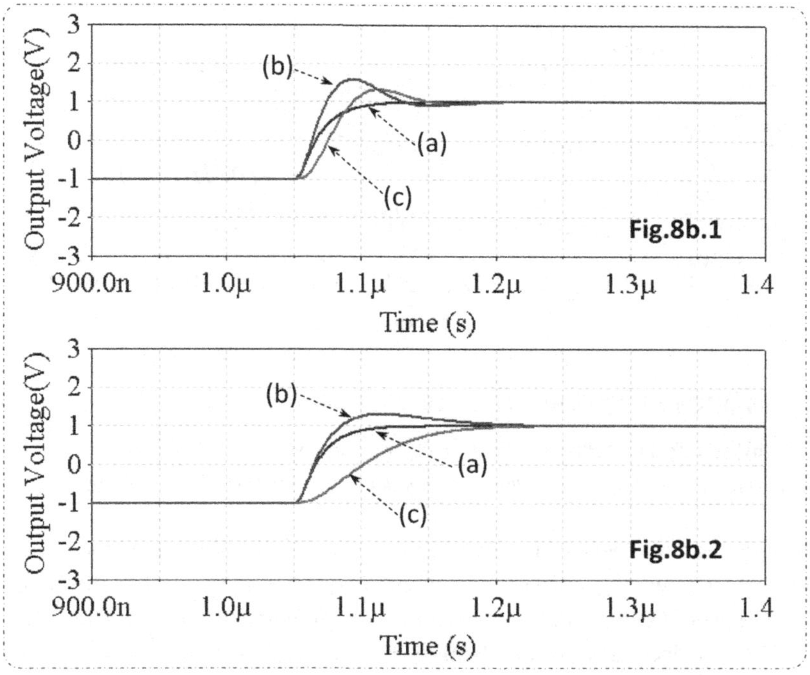

Fig. 8b.1 $f_{U2}=f_{C1}$ **=10MHz:**
(a) target amplifier (A_1)
(b) AEF EC amplifier ($F_{EC}A_1$) of Fig.6
(c) FDR EC amplifier ($F_{EC}A_1$) of Fig.7

Fig.8b.2 $f_{U2}=0.3f_{C1}$ **=3MHz:**
(a) target amplifier (A_1)
(b) AEF EC amplifier ($F_{EC}A_1$) of Fig.6
(c) FDR EC amplifier ($F_{EC}A_1$) of Fig.7

Figure 8b Step response of amplifiers in Fig.6 (AEF) and Fig.7 (FDR).

Fig.9 illustrates the error correction performance versus frequency for both FDR and AEF schemes, in the same two specific situations of Fig.8.2: $f_{U2}=f_{C1}$=10MHz and $f_{U2}=0.3f_{C1}$=3MHz. These curves confirm that reducing fU2 for improving system stability margins will proportionally reduce the error-correction performance in both schemes.

AEF and FDR review conclusions. Macdonald's AEF scheme was assessed in comparison with FDR, the standard feedback correction scheme, in the classical application where the target amplifier is the output stage of a power amplifier, with nominal unity gain. We have seen that the two systems, although operating in dissimilar ways, share some important features. They have exactly the same stability criteria and the same distortion reduction ratio. So when $f_{U2}=f_{C1}$ both have the same stability margins and the same potential of distortion reduction ratio ρ_P for both V_{e1} and V_{e2} errors shown in Fig.6 and Fig.7.

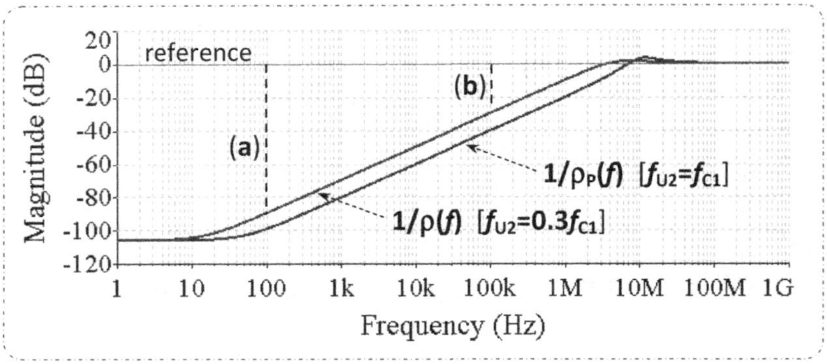

(a) $|\rho(100\text{Hz})|$: 100dB with $f_{U2}=f_{C1}$=10MHz; 90dB with $f_{U2}=0.3f_{C1}$=3MHz

(b) $|\rho(100\text{kHz})|$: 40dB with $f_{U2}=f_{C1}$=10MHz; 30dB with $f_{U2}=0.3f_{C1}$=3MHz

Figure 9 Inverse of the error/distortion cancellation ratio ρ(f) of FDR and AEF schemes.

None of the two systems meets the fourth requirement for EC schemes, in particular when $f_{U2}=f_{C1}$, and therefore they show substantially different time-frequency responses than the target amplifier to be corrected, A_{1T}. The impact of FDR on A_{1T} transfer function, however, is higher than AEF for $f>f_N$ where FDR gain decreases, independently on the ratio f_{U2}/f_{C1}, at a rate of -40dB/dec, while the gain of A_{1T} decreases at a more comfortable and manageable rate of -20dB/dec. So the FDR corrective potential in general cannot be fully exploited for reasons of system stability, especially when FDR is used inside the overall feedback loop of the audio amplifier, where it is more difficult to protect the stability margins of the system and it is in general necessary to reduce the f_{U2}/f_{C1} ratio.

AEF shows a better frequency-response for $f>f_N$, as its high frequency gain decreases at a rate of -20dB/dec, as A_{1T}. This key feature bodes well for applications where AEF scheme is used within the overall feedback loop of audio power amplifiers, since system stability can be secured with more favorable values of f_{U2}/f_{C1} ratio than FDR. Therefore higher ratio of error correction (but generally not the full AEF potential!) than FDR may be achieved. However, AEF shows worse step-response than FDR, since overshoot and settling time are strictly intertwined together, and finding a good trade off in critical applications (e.g. video amplifiers) might be troublesome.

Hawksford Feedback Error Correction (HFEC) [FBEC option (II)]

The second FBEC solution I will review now is the HFEC scheme presented in its conceptual form by M. J. Hawksford in 1981 [1]. The HFEC basic block diagram is depicted in **Fig. 10.** It has graphically the same number of blocks as AEF: two voltage summers, **S3** and **S4**, and the block $\mathbf{A_e}$, which in this context has the function of buffering the error voltage $\mathbf{V_e}$ and has nominally unity gain.

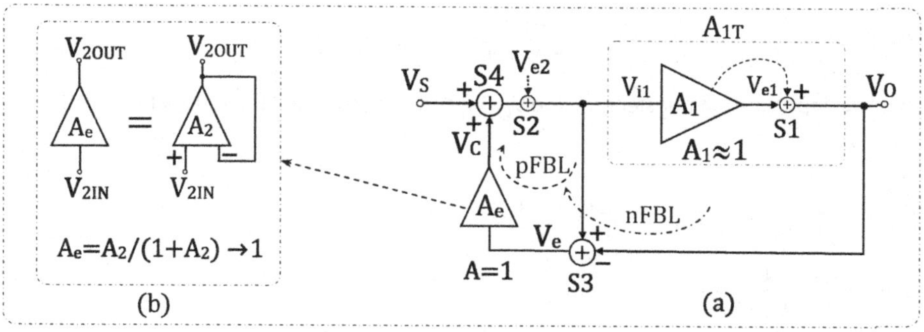

Figure 10 Block diagram of the Hawksford Feedback Error Correction scheme (HFEC).

Often $\mathbf{A_e}$ is not shown in the HFEC block diagram and is replaced by a simple connection, but this gained graphical simplicity can be misleading. $\mathbf{V_{e2}}$, as in AEF and FDR, represents the overall error (**noise, offset, distortion, etc.**) associated with the added blocks and is of course part of the game. Yet, there are some important and clear differences with the AEF scheme to highlight:

- the error in the HEFC scheme is the difference between the output Voltage $\mathbf{V_o}$ and the input voltage $\mathbf{V_{i1}}$ of the target amplifier, and not the difference between $\mathbf{V_o}$ and the input signal voltage $\mathbf{V_s}$ as in AEF;
- the mission of generating the necessarily accurate error correction voltage $\mathbf{V_C}$ is not entrusted to the automatic mechanism of a feedback loop and consequently to the accuracy-precision of its high gain amplifier, as occurs in the AEF scheme, but is appointed to a specific set of functional blocks which have the challenging task to produce a perfect copy of the error to be corrected, $\mathbf{V_{e1}}$, at all frequencies and voltage levels;
- the corrective voltage $\mathbf{V_C}$ is applied upstream of the sensed reference voltage, in this case $\mathbf{V_{i1}}$;
- but the most intriguing difference with AEF is that apparently in the HFEC scheme there is no need for any EC loop gain, since block $\mathbf{A_e}$ has unity gain (which may sound and look simple to implement !), and all voltage summers, as usual, have unity gain for all their inputs. But we must not be misled by the disarming simplicity of the basic HFEC scheme. All the voltage gain needed for the system to work properly is provided by the regenerative process associated with the embedded positive feedback loop (pFBL) which is the heart (as well as the *gain engine*!) of the scheme (see J. Didden [9]);
- the combination of A_e and S3 is functionally equivalent to a difference amplifier with unity gain, while in AEF the cascade of S3 and A_2 blocks, is functionally equivalent to a high gain Op-Amp.

The circuit operation is apparently simple. First, the error voltage $\mathbf{V_e}=\mathbf{V_{i1}}-\mathbf{V_O}$ (which, again, contains both linear and non-linear components) is sensed and extracted from the target amplifier $\mathbf{A_1}$, by means of the summer **S3**. An exact copy of it, i.e. $\mathbf{V_C}\cong\mathbf{V_e}$, is then provided by the block $\mathbf{A_e}$ to the input of summer **S4**, which adds this correction voltage to the input signal voltage $\mathbf{V_S}$, as occurs in all FBEC schemes. The resulting voltage at the input of the amplifier to be corrected, A_1, is then given by $\mathbf{V_{i1}}=\mathbf{V_S}+\mathbf{V_C}+\mathbf{V_{e2}}$. The output voltage can then be easily calculated and is given by

$$V_o(HFEC) = V_s A_1 \frac{1}{1 - A_e + A_1 A_e} + V_{e1} \frac{1 - A_e}{1 - A_e + A_1 A_e} + V_{e2} \frac{A_1}{1 - A_e + A_1 A_e} \tag{19}$$

Therefore, the input-output transfer function V_O/V_S and our equivalent EC transfer function F_{EC}, can be written as

$$\frac{V_o}{V_s}(HFEC) = A_1 \frac{1}{1 - A_e + A_1 A_e} = A_1 F_{EC}(HFEC) \tag{20}$$

$$F_{EC}(HFEC) = \frac{1}{1 - A_e + A_1 A_e} \tag{21}$$

The relevant stability requirement for the HFEC scheme, to be met in all operating conditions, is given by

$$1 - A_e + A_1 A_e \neq 0 \tag{22a}$$

In our context, i.e. A_e having constant gain (close to 1) as well as infinite bandwidth and $A_1 = A_{10}/(1+s/p_1)$, with $A_{10} \cong 1$, inequality (22a) turns into

$$[1 + A_e(A_{10} - 1)]/[1 - A_e] \geq 0 \tag{22b}$$

which, for values of A_{10} close to 1 as in our case, implies that stability is granted only if $A_e \leq 1$.
Eq.19 also provides information about the residual HFEC output error V_{eo}, which is given by

$$V_{eo} = V_{e1}/\rho_1 + V_{e2}/\rho_2 \tag{23}$$

where $\rho_1 = (1 - A_e + A_1 A_e)/(1 - A_e)$ and $\rho_2 = (1 - A_e + A_1 A_e)/A_1$ are the **HFEC error cancellation ratios** for V_{e1} and V_{e2}, respectively.

The ideal situation occurs of course when $A_e \to 1$ (at all frequencies and voltage levels, yet without exceeding unity!), so we have

$$\lim_{A_e \to 1} V_o(HFEC) = V_s + V_{e2} \tag{24}$$

and

$$\lim_{A_e \to 1} \rho_1 = A_1/(1 - A_e) \to \infty \tag{24a}$$

$$\lim_{A_e \to 1} \rho_2 = A_e \to 1 \tag{24b}$$

The above equations explain the well-known features of the ideal HFEC scheme: regardless of the transfer function of the target amplifier, when $A_e = 1$ at all frequencies (i.e. infinite bandwidth!) and in the entire operating range (i.e. unlimited dynamics and perfect linearity!), the HFEC output voltage V_o is ***cogently*** equal to the input voltage V_s at all frequencies and at all levels, and the error of the target amplifier A_{1T} is completely **cancelled**. The error correction circuit will provide to A_{1T} the instantaneous input voltage it needs (if its transfer function is not zero, otherwise the input voltage would need to approach infinity!) to compel the output voltage of this ideal correction system, to be always exactly equal to the input voltage. But if we move slightly away from this ideal and limit situation, for instance by slightly reducing the bandwidth of A_e to, say, a hundred times the corner frequency of the target amplifier A_1, then the HFEC system will become unstable (or very critically stable). So, for all practical HFEC implementations, as for AEF, there is an upper limit not to be exceeded for A_e bandwidth, if we want the system to operate with good stability margins. This issue will also be part of the discussion which follows.

I think it is appropriate at this point to examine in more detail the implications of all the above equations and highlight the following points:

- to ensure stability and meet the condition given by eq.(22), we must always have $\mathbf{A_e \leq 1}$;
- if $A_e>1$ our system (where $A_{10} \cong 1$) will be intrinsically unstable;
- on the other hand, the error V_{e1} of the target amplifier can be reduced to a large extent only if A_e is very close to unity (so that $\rho_1 \rightarrow \infty$) ;
- so we have two conflicting needs to balance, that can be concisely written as $\mathbf{1 > A_e \rightarrow 1}$;
- a **careful calibration** (gain)**, equalization** (vs frequency) **and stabilization** (against gain variations in the operating range, temperature variations, ageing, etc) is then needed to optimize HFEC error correction performance. **This is the first major limit of HFEC;**
- if for instance $A_e=0.9$ and $\rho_1=10$, V_{e1} will be reduced not more than 20 dB. To achieve an error reduction better than 60dB at all useful frequencies, output voltage levels and operating conditions of the target amplifier, A_e must necessarily be critically comprised in the following range: $0.999<A_e<1$;
- in addition, even in the ideal conditions of $A_e=1$, the HFEC is completely ineffective against the error $\mathbf{V_{e2}}$ produced by the same EC circuit blocks (S3, S4 and A_e), that have been introduced into this context just to cancel the error of the target amplifier. In fact, since $\rho_2 \rightarrow \mathbf{1}$, this error will emerge virtually unaltered at the output of the corrected amplifier and could even dominate the residual output error $\mathbf{V_{eo}}$. **This is the second major limit of HFEC;**
- so, for the HFEC system to be truly effective, the error associated with the actual HFEC circuit needs to be much lower than the error of the target amplifier, that is $V_{e2} \ll V_{e1}$.

In order to overcome all above limitations (intrinsic error, accuracy and stability, etc.), and let the HFEC scheme express its entire corrective potential, in my opinion there is no alternative: **a good amount of ordinary negative feedback is to be used inside the HFEC circuit,** in conjunction **with high precision active and passive components,** and more specifically very low tolerance and low temperature coefficient (TC) resistors!

An effective way of introducing negative feedback into the HFEC scheme in order to exploit its corrective potential without putting circuit stability in jeopardy, as the simplest example that comes to mind, is depicted in **Fig.10(b)**, which shows a possible and practical implementation of block A_e, which employs a high dc gain and precision op-amp $\mathbf{A_2}$ (e.g. the same single pole Op-Amp used in AEF) and negative feedback to form a voltage follower with the needed gain accuracy and intrinsically low levels of error. I will use here the same terminology as in AEF and FDR, in order to facilitate the comparison between these different schemes. If we then substitute the gain formula shown in Fig.10(b), i.e. $A_e=A_2/(1+A_2)$, into eq. (19) and develop, we obtain

$$\mathbf{V}_o(HFEC) = V_s A_1 \frac{1+A_2}{1+A_1A_2} + V_{e1}\frac{1}{1+A_1A_2} + V_{e2}\frac{A_1(1+A_2)}{1+A_1A_2} \quad (25)$$

which is **surprisingly** the same as the corresponding eq.(5b) of the AEF scheme, the only difference being the term including $\mathbf{V_{e2}}$, which in AEF is **much lower** and equal to $\mathbf{V_{e2}}A_1/(1+\mathbf{A_1A_2})$. This means that, apart from the third term on the right side of eq.(25), which deserves some specific considerations to be made later on, all the points made regarding AEF can be directly applied to the HFEC scheme too. In particular the analysis on the time- frequency response, the stability criteria (**the simple condition $A_e \le 1$ no longer applies**) and the error correction potential, is fully valid for HFEC, too. So the input-output and EC transfer functions can be re-written as

$$\frac{\mathbf{V}_o}{V_s}(HFEC) = \frac{\mathbf{V}_o}{V_s}(AEF) = A_1\frac{1+A_2}{1+A_1A_2} \quad (26)$$

$$F_{EC}(HFEC) = F_{EC}(AEF) = \frac{1+A_2}{1+A_1A_2} \quad (27)$$

The residual output error $\mathbf{V_{eo}}$ is instead different from AEF. It can be written once again as

$$\mathbf{V}_{eo} = V_{e1}/\rho_1 + V_{e2}/\rho_2 \quad (28)$$

where ρ_1 and ρ_2 are the usual **error cancellation ratios** for $\mathbf{V_{e1}}$ e $\mathbf{V_{e2}}$, respectively, but while $\rho_1=(1+\mathbf{A_1A_2})$ is the same as for AEF, $\rho_2=(1+\mathbf{A_1A_2})/[\mathbf{A_1}(1+\mathbf{A_2})]$ is completely different from its AEF counterpart.

In HFEC in fact, since $A_1 \cong 1$, for $A_2 >> 1$ we have

$$\rho_1 \cong A_2 \quad (29a)$$

and

$$\rho_2 \cong 1 \quad (29b)$$

Due to the perfect equivalence of HFEC with AEF in terms of system dynamics and error correction performance, I will not repeat the detailed analysis I made for AEF, since the results reached for AEF are exactly the same for HFEC. Of course all conclusions on time-frequency response, stability criteria and on the optimal f_{U2}/f_{C1} ratio for AEF, as well as the simulation results shown in Fig.8a-8b and Fig.9, are directly and immediately applicable to HFEC.

At this point I have to stress that these results are valid not only for the example of Fig.10(b) taken into consideration, but for all possible implementations of HFEC (included those ones incorporating one auxiliary AEF loop **[2]** and having similar time-frequency behavior and stability criteria to plain AEF or HFEC). We could for instance consider a second example where A_e consists of a simple RC low pass filter (but beware, this example cannot be looked at as a possible implementation!) with transfer function equal to $\mathbf{A_e} = \mathbf{A_{eo}}/(1+s/\mathbf{p_{u2}})$ (with $A_{eo}=1$), and where the target amplifier A_{1T}, as in AEF, has the usual first order transfer function equal to $\mathbf{A_1}=1/(1+s/\mathbf{p_1})$. If we substitute, for instance, in eq. (21) and develop, we achieve

$$F_{EC}(HFEC) = \frac{1}{1 - A_e + A_1 A_e} = \frac{(1 + s/p_1)(1 + s/p_{U2})}{1 + s/p_{U2} + s^2/(p_1 p_{U2})} = \frac{1 + s(p_1 + p_{U2})/\omega_N^2 + s^2/\omega_N^2}{1 + s(2\zeta/\omega_N) + s^2/\omega_N^2} \quad (29c)$$

which is exactly equal, again, to the corresponding eq. (14b) of AEF scheme.

Some additional considerations are instead worth to be made and highlighted about the implications of the ideal and limit case of $A_e=1$ in the HFEC scheme. If we consider the implementation example of block A_e proposed in Fig. 10(b) (but the same applies to other possible implementations as well) we should note that the limit case of $A_e \to 1$ converts into the limit case of $A_2 \to \infty$ in Fig. 10(b). If we apply the same limit of $A_2 \to \infty$ to AEF and FDR schemes of Fig. 6 and Fig.7, we achieve the same limit results of HFEC where V_o is identically and **cogently** equal to V_s at all levels and frequencies, independent from the real characteristics and performance of the target amplifier to be corrected A_{1T}, and the residue of error V_{e1} is zero at the output of the corrected system. So, the apparently inoffensive assumption of $A_e=1$ at all frequencies made for the HFEC block diagram, is in truth hiding the much more demanding assumption of a gain tending somewhere to infinity (at all frequencies!). If we consider instead more realistic assumptions for A_e, if we assume for instance that A_e has exactly unity gain but in a limited bandwidth, than we have seen that the HFEC dynamic performance becomes exactly equivalent to the more familiar AEF scheme, yet HFEC capability to correct the error V_{e1} of the target amplifier, can approach but never exceed AEF performance, as shown in **Fig. 11** by the simulation results of the inverse of the error cancellation ratio $\mathbf{1/\rho_1}(f)$ achievable by HFEC with different gain values of $A_e = A_2/(1+A_2)$ in Fig.10. We can appreciate that the best achievable HFEC result is the same as AEF.

Finally, a few more words on the error cancellation ratio ρ_2 of the error produced by the EC circuit. Eq. (29b) tells us that HFEC, contrary to AEF and FDR, is ineffective in reducing V_{e2} even in the ideal case of $A_e=1$ at all frequencies [i.e. $A_2=\infty$ in Fig.10(b)], so it is extremely important that designers are fully aware of this limit and take the necessary countermeasures in the design and implementation phases. I would like to emphasize here, in particular, two simple rules:

- A_2 should be a high-precision, low noise and low distortion Op-Amp (as in AEF) with $f_{U2} > f_{C1}$;
- Voltage summer S4 should, to the maximum possible extent, be implemented with passive components only or alternatively with active devices (Op-Amps included) having, in addition to high-precision, low noise and low distortion, a much larger bandwidth than the target amplifier A_{1T}, in order to allow the full error correction potential of HFEC to be exploited.

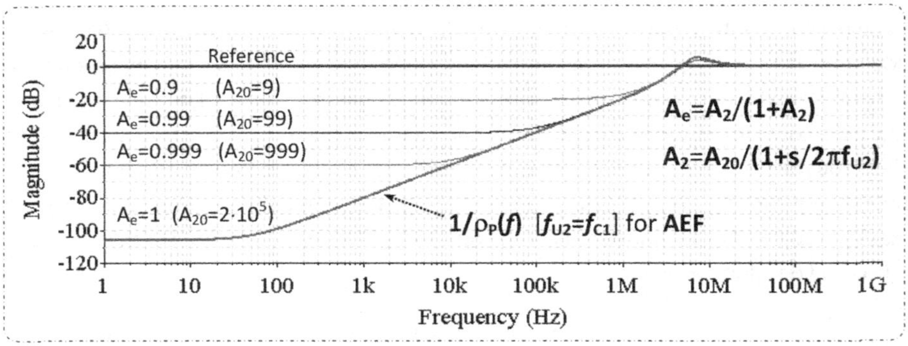

Figure 11 *Inverse of HFEC error- cancellation ratio* $1/\rho_1(f)=V_{eo}/V_{e1}$ *for* $f_{U2}=f_{C1}$ *and different values of* A_e

HFEC review conclusions. HFEC has been assessed in the same application as AEF , where the target of the EC process is the output stage of a power amplifier, with nominal unity gain. The results of this assessment are that HFEC systems have the same input-output transfer function, the same F_{EC} transfer function, the same stability criteria and, **when perfectly calibrated**, the same error correction capabilities as AEF scheme. Therefore the conclusions reached in the discussion of the AEF are valid also for HFEC. However, we have also seen that HFEC has some distinctive limitations which are **completely absent in the AEF scheme**:

- the need for a **careful calibration, equalization and stabilization** to ensure that A_e=1 in order to develop its full error correction potential (same levels of ρ_1 as AEF);
- the need for **very low levels of intrinsic error** (distortion, noise, offset, etc) since HFEC is completely ineffective in reducing V_{e2} [$\rho_2 \cong 1$] and therefore V_{e2} will re-appear virtually unchanged at the output of the system.

We have seen that a good way of mitigating the effects of these limitations is to use ordinary negative (local) feedback inside the HFEC circuit as well as high precision active and passive components. However, the problem of the potentially higher residual error of HFEC compared with AEF, remains unsolved.

In conclusion my opinion is that if negative feedback has to be used inside the HFEC scheme to get closer to the error correction performance of AEF, then the best design choice, **in general,** is to employ the **AEF scheme directly, since it makes the best possible use of the available excess loop gain (A_2), with virtually the same efficacy against both V_{e1} and V_{e2}.** Moreover, this result is achieved without increasing circuit complexity and without introducing the extra cost associated with the critical selection of components and/or calibration procedure HFEC needs to get closer to AEF error correction performance. I said AEF *in general* is a best choice than FFEC. However this is not always true, since there are some good exceptions: in some applications with less stringent distortion requirements HFEC solutions based on some commercial wide bandwidth and low distortion active devices [e.g. Current Feedback (CF) Op-Amps], may be easier and less expensive to implement than AEF.

Example of implementation of AEF and HFEC schemes

I think it is now worthwhile to examine and compare some simple and practical examples of implementation of both AEF and HFEC error correction schemes that have been discussed so far, which use the same active devices and come very close to fully exploit their respective error correction potential ρ_p, when applied to the output stage of a power amplifier. But before going into details, I would like to spend a few words about the decision process steps I apply in the design of these low distortion amplifiers with error correction, which can be summarized as follows:

1. definition of requirements of the error correction circuit (bandwidth, voltage and current dynamic range, stability, etc) based on the characteristics of the target amplifier to be corrected, and available supply voltages;
2. selection of the active devices to be used;
3. choice of the implementation architecture;
4. validation of the chosen architecture and devices;
5. detailed design of the implementation.

So the first thing I do is the selection of the power devices to be employed in the design. MOSFETs are my preferred power devices for several reasons: they are rather cheap nowadays, robust, easy to drive and, one of the most important features in our context, they have intrinsically very wide bandwidth (more than 10MHz can be achieved when driven with equivalent gate resistance of 20Ω in a wide output voltage excursion, say up to about 80-90% of the supply rails) which allows to achieve very high error correction ratios in a wide frequency range, since $|\rho_p| \cong f_{C1}/f$ [see eq.(18)].

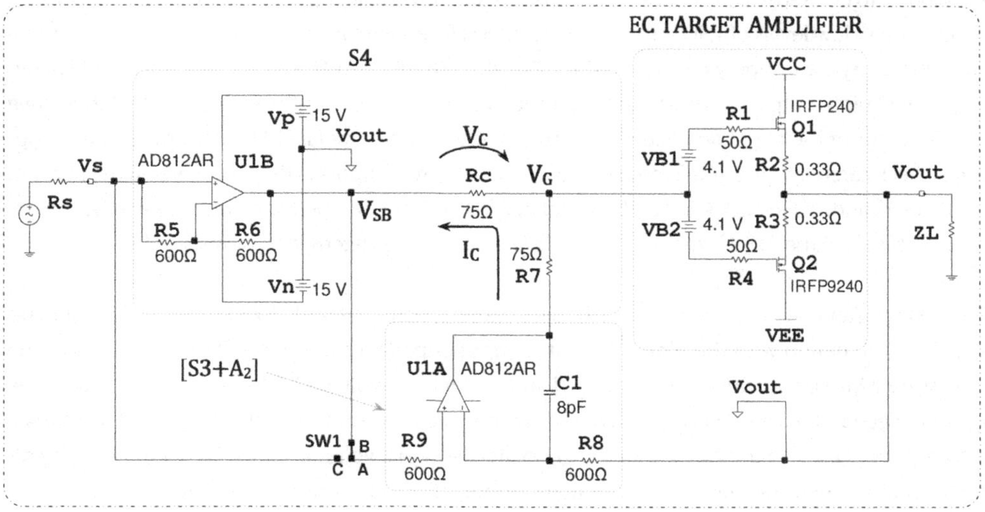

Figure 12 A practical implementation of AEF applied to a MOSFET power output stage

Let's start with AEF then, and refer to **Fig.12** which shows a simplified circuit diagram of the implementation example I will discuss (details of protections, floating power supply and MOSFET biasing circuits, load decoupling filter, etc., are not shown). The target amplifier is composed of an N&P pair of power MOSFETs arranged in a source follower configuration.

Supply voltage is assumed to nominally be ± 55V in order to provide an undistorted voltage output swing of about $90V_{P-P}$ on an 8Ω load, although it can swing very close to the supply rails and have a clean clipping behavior. MOSFET bias current is set to I_{DQ}≅100mA via split bias voltage (VB1 and VB2), which provides a conveniently centered (close to V_{OUT}) single drive-point for the target amplifier. When driven with less than 50Ω equivalent gate resistance its corner frequency is f_{C1}(-3dB)≥10MHz, its voltage gain is about 0.92, and its actual transfer function is single pole, therefore ideal for EC application. To provide an output swing of $90V_{P-P}$ up to 20 kHz it typically needs a gate drive voltage $\mathbf{V_G}$ of about 97 V_{P-P}, and therefore a (V_G-V_{OUT}) drive voltage of 7 V_{P-P}. It means that if I want to use commercial ± 15V op-amps for the EC circuit I need to refer the central point of their dual power supply (V_{REF}) not to ground but to the output voltage (power supply bootstrap) and make sure this floating circuit can tolerate this extra positive feedback loop.

In general this technique entails for the bootstrapped circuit, high levels of high-frequency power supply rejection ratio (PSRR) on both supply rails, which is not common in ordinary op-amps, the PSRR of which is not adequate and you usually need to reduce f_{U2}/f_{C1} ratio to achieve the needed stability margins of the EC loop; but this , as we know, will proportionally reduce the distortion cancellation ratio. Current feedback (CF) op-amps with a single voltage gain stage architecture are usually good candidates to be used in this applications. The dual current-feedback op-amp used in Fig.12, the AD812, in addition to low levels of THD (<0.002% up to 100kHz) also with load resistance as low as 150Ω, features in fact a power supply rejection ratio higher than 30 dB up to 10MHz, as well as the wide bandwidth and high output drive capabilities (both voltage and current) needed to meet the drive requirements of the target amplifier. By the way, if you want to simulate PSRR performance of an op-amp, check carefully if the SPICE model used by your EDA tool supports this parameter in the relevant frequency band. In Fig.12, S3 and A_2 blocks of AEF in Fig. 6, are grouped and implemented by the op-amp section A of the AD812, the open-loop dc gain of which is determined by resistor R8.

The voltage summer S4 is implemented by a low value resistor network (R7≅ RC≅75Ω) in order to retain, in combination with the highly capacitive input impedance of the MOSFET output stage, the necessary wide bandwidth [f_C(S4+A_1)≅f_{C1}] and a wideband voltage buffer, implemented with the op-amp section B of the AD812, to achieve the intrinsic input-output isolation of the S4 summer, which in this case is mandatory in order to isolate the signal voltage source **Vs** from the high correction current $\mathbf{I_C}$ (about 46mA$_P$) which flows through R_C and is filled up with all the harmonics of the input signal. Many well-known EC implementations seem to overlook this very harmful side-effect and lose a good part of the benefits introduced by EC.

The ratio R7/Rc, in conjunction with the maximum output voltage and current of U1A, determines the maximum level of achievable corrective voltage V_C. The values shown are compatible with the needed corrective voltage of about $3.5V_P$ (both linear and non-linear error are included) and relevant output current of $46mA_P$. Capacitance C1 can be adjusted to achieve the required stability margins of the error correction loop [$f_{U2} \cong f_C(S4+A_1) \cong f_{C1}$]. With the values shown in Fig.12, the error reduction is higher than 35dB up to 100 kHz.

Simulated THD (15 harmonics included) at V_{out}= $45V_P$ on a load ZL=8Ω, with SW1 in AB position, is less than 0.0001% at 200Hz, about 0.0002% at 2kHz, and lower than 0.002% at 20kHz. Since the distortion of the target amplifier at V_{OUT}=$45V_P$ is about 0.2% at all audio frequencies, the achieved improvement is always higher than 40dB, and this improvement is constant with lower levels of the input signal V_S. Moreover, these results are valid for any source resistance $\mathbf{R_s}$ lower than or equal to 10kΩ. This feature widens the application range of this AEF solution. SW1 switch is used only to indicate that either the input signal V_S (pos. AC) or its buffered copy V_{SB} (pos. AB) can be used as the AEF input reference voltage, without large differences in performance. The design choice may depend on the utilization context. In pos. AB the stability margins are slightly better when this error-corrected output stage is used inside the overall feedback loop of a complete power amplifier, but distortion is slightly higher, due to the contribution of the input buffer which is left outside the EC loop. Slew rate and power bandwidth (with protections not active!) with R_s<1kΩ, are SR>±400V/μs and 800 kHz (yet only for short times not to exceed the maximum junction temperature of the AD812), respectively.

In **Fig.13a** we can see the implementation example of the HFEC scheme which employs the same active devices and part of the architecture of the AEF implementation example shown in Fig.12.

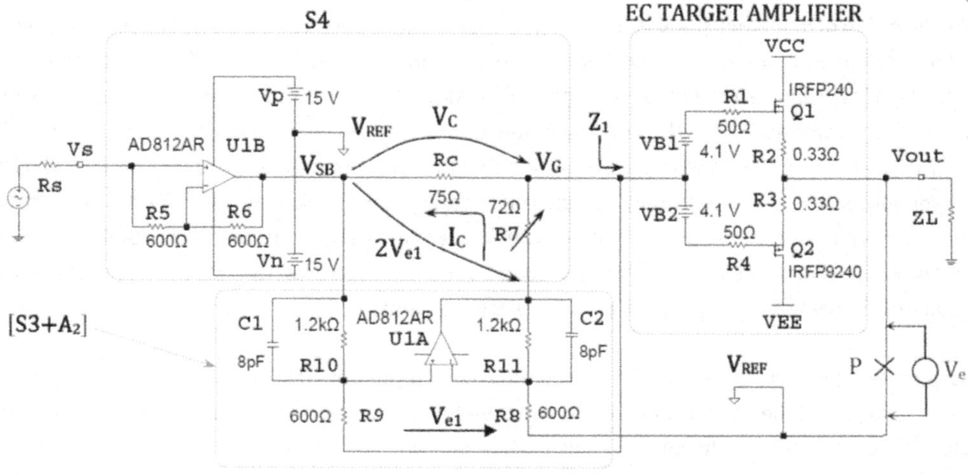

Figure 13a A possible implementation of HFEC applied to a MOSFET power output stage.

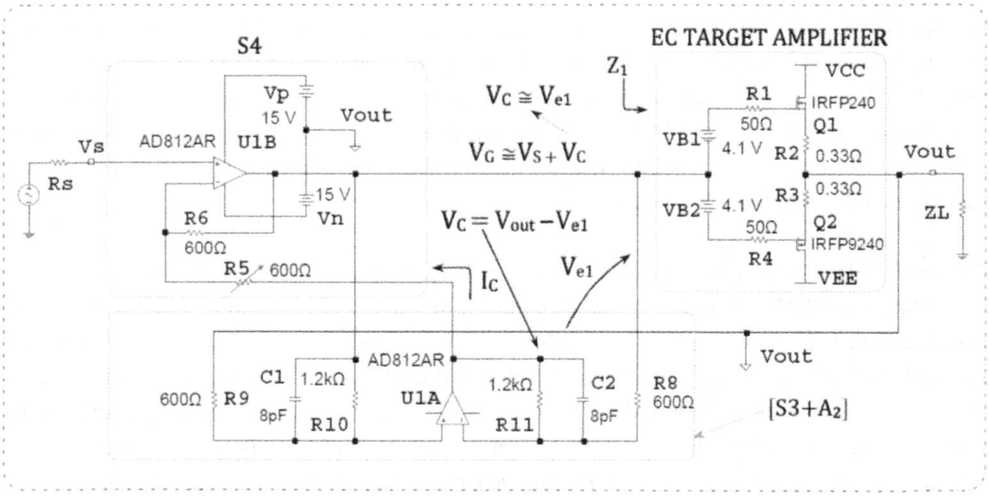

Figure 13b An improved implementation of HFEC applied to a MOSFET power output stage.

The difference is that blocks S3 and A_e of Fig.10 are grouped and implemented by the difference amplifier with gain of 2 built around the op-amp section A of the AD812 with resistors R8, R9, R10 and R11. The difference amplifier has a gain of 2 to compensate for the attenuation of the summing resistor network that has a nominal attenuation of 2. Capacitors C1 and C2 can be adjusted to achieve the required stability margins of the error correction loop ($f_{U2} \cong f_{C1}$). Calibration of the EC circuit gain ($A_e \rightarrow 1$) is done by adjusting R7 (its nominal value is 72Ω). In this way it is possible to achieve an error correction ratio of about 60dB at low frequency and 35dB at 100 kHz, which is virtually the same as for AEF. With the values shown in Fig.13a, simulated THD (15 harmonics included) at V_{out}= 45V_P on a load ZL=8Ω, **is less than 0.0004% at 200Hz, about 0.0034% at 2kHz, and 0.034% at 20kHz.** Since the distortion of the target amplifier is about 0.2% at all audio frequencies, the achieved improvement is higher than 35dB up to 2kHz, but not more than 16 dB at 20kHz, that is 24 dB lower if compared with the 40dB improvement obtained with AEF at 20kHz. Slew rate and power bandwidth are substantially the same as AEF. This large THD difference at the highest audio frequencies is due to the different sensitivities of AEF and HFEC schemes to the non-constant input capacitance of the MOSFET power output stage (which strongly depends on the drain-source voltage of MOSFET devices and on their forward trans-conductance variations with V_{OUT}), and determines the quality of its input impedance, Z_1 in Fig. 13a. With output stages using bipolar transistors we need to take into account the additional impact on Z_1 of β variation with collector current.

As a matter of fact, in Fig. 12 and Fig.13, the equivalent input impedance Z_1 of the output stage, is an integral part of the voltage summer S4, and therefore the corrective voltage V_C (**which is outside the HFEC correction loop**) will also depend on Z_1 according to the relationship $V_C=V_{e1}/(1+R_7/2Z_1) \neq V_{e1}$. Therefore, since Z_1 depends on the output voltage V_{out}, V_C can never be (even after calibration) an exact copy of V_{e1} at all frequencies and signal levels, and Z_1 will then become a major contributor of the error [$V_{e2}(Z_1)$] which cannot be counteracted by the HFEC loop. In AEF, on the contrary, sensitivity to Z_1 variations with the output voltage V_{out} is much lower than in HFEC, because the associated source of error, $V_{e2}(Z_1)$, is fully embraced by the AEF EC feedback loop, and can therefore be consistently reduced to the expected low value, at all frequencies and signal levels.

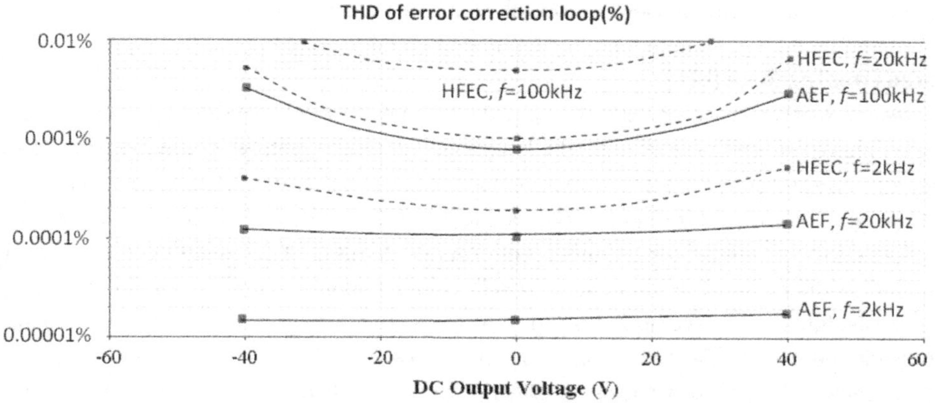

Figure 14 Harmonic Distortion (THD) of the Error Correction Loop (ECL) for AEF and HFEC implementations of Figure 12 and Figure13a. Number of harmonics, nh=16; level of ac Error Voltage V_e=1V_P

A practical and effective way to assess the quality and the efficacy of the chosen AEF or HFEC implementation, is to simulate (and possibly measure), in addition to the actual error cancellation ratio, the THD performance of the error correction (EC) loop by adding to the output (point P in Fig.12 and Fig.13a) an ac error (V_e) of fixed amplitude (say 1V_P) and by simultaneously applying at the input a variable dc voltage ($V_{s(DC)}$) comprised in the specified output voltage swing, which in this case is equal to ±45V. **Fig.14** shows the simulated THD performance of AEF and HFEC EC loops in the implementations of Fig. 12 and Fig.13a at different frequencies, as a function of the corresponding dc output voltage. These simulation results confirm that THD of the AEF EC loop is more than 20 dB better than HFEC for dc output voltages up to ±40V even at frequencies as low as 2 kHz.

Overall distortion performance of the HFEC implementation of Fig.13a can be substantially improved by using the U1B CF Op-Amp of the AD812 not only for buffering the input voltage but also for the S4 summing function as shown in **Fig.13b** (which incidentally, among my HFEC implementations, is the preferred one for its compactness and low IC/chip-component count). In this solution the contribution by the non-constant input capacitance of the power MOSFET output stage to the overall amplifier distortion, left outside the HFEC correction loop in Fig.13a, is virtually eliminated, and HFEC can almost reach the distortion performance of its AEF counterpart. Of course a careful calibration (e.g. via R_5) as well as a very low temperature- coefficient (TC) resistor are still needed, but in virtue of the low intrinsic distortion of the AD812 and its very large bandwidth (more than 100MHz) the nominal THD performance of HFEC implementation of Fig.13b, with V_{out}= 45V_P on a load ZL=8Ω, turns out to be less than 0.0001% at 200Hz, about 0.0002% at 2kHz, and lower than 0.002% at 20kHz, more or less the same as AEF. By the way, the most compact HFEC solution, to my best knowledge, was invented by Bruce Halcro Candy, US patent N. 6052027 **[9].** It employs in a very effective way, a single Op-Amp to implement the difference amplifier and the voltage summer and is currently used in commercial products. In my opinion, however, it needs an input buffer to reach its full error-correction potential, so in terms of overall compactness it is not so different from Fig. 13b.

Balanced Error Feedback (BEF)

In the previous sections I have examined two popular schemes of error and distortion reduction, AEF and HFEC. I have proven that AEF and HFEC have a few important features in common: time-frequency response and error correction potential. However, amplifiers using AEF show in general better THD performance than those using HFEC, due to insufficient calibration and/or the inability of HFEC to effectively counteract some sources of error (V_{e2}) internal to HFEC circuitry. Compared with the ordinary feedback loop (FDR), both AEF and HFEC have the additional property of a lower impact on the transfer function of the amplifier being corrected, so they are closer than FDR in meeting the fourth requirement for the ideal error correction scheme (i.e. F_{EC}=1). Therefore they can be used inside the overall feedback loop of a power amplifier with less stability problems than FDR. This impact can be further reduced by lowering the ratio f_{U2}/f_{C1}, but with the penalty that also the error cancellation ratio will be proportionally reduced.

So the consequent question we can now ask ourselves is whether it is possible for an FBEC scheme to fully match the fourth requirement, but without limiting its error correction capability. A positive answer to this question is offered by the **Balanced Error Feedback** scheme shown in **Fig.15 [12]**, applied to our typical target amplifier $\mathbf{A_{1T}}$ with nominal unity gain, that is $A_1 \approx 1$ (e. g. the output stage of an audio power amplifier). The novelty, compared with AEF and HFEC, is that here the input signal $\mathbf{V_s}$ **is routed on two similar parallel paths**, the main path w_M and the **balancing** path $\mathbf{w_B}$, through which they reach the input terminals of the error amplifier A_2. The outstanding features of this new scheme are essentially based on two facts:

- the balancing block $\mathbf{A_B}$ has in general unity gain and no load at its output (only the high input impedance of A_2), so it may consist of **passive components only** and is virtually **free of non-linear distortion;**
- if the two paths, w_M and w_B, have exactly the **same gain and time-frequency response**, i.e. they are well balanced, the main signal component ($\mathbf{V_s}$) will appear at the inputs of A_2 **as a common mode voltage only**, and the difference $\mathbf{V_d = V_b - V_m}$ will contain **only the components of the errors to cancel ($\mathbf{V_{e1}}$ and $\mathbf{V_{e2}}$).**

The expected result, **if $\mathbf{A_2}$ has adequate CMRR and input common mode voltage range**, will be that the transfer function and the dynamic properties of **the corrected system will be exactly the same as the non-corrected amplifier $\mathbf{A_{1T}}$, so that $\mathbf{F_{EC} = F_{ECO} = 1}$ and the forth requirement will be fully met, independently of the f_{U2}/f_{C1} ratio.**

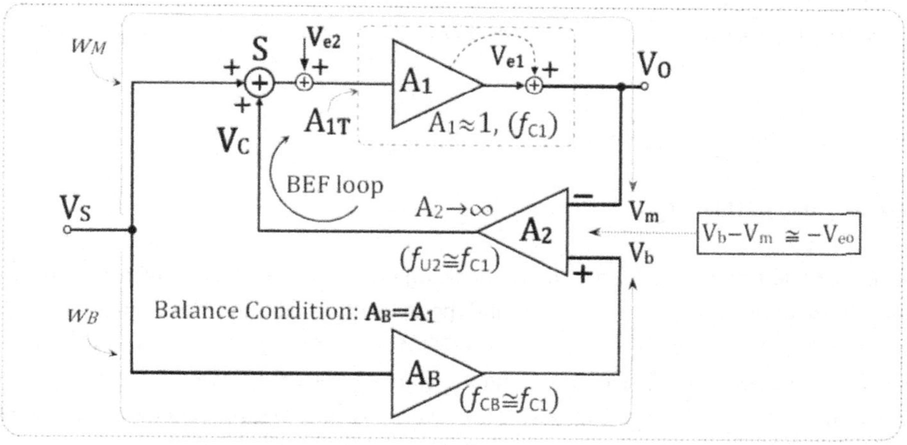

Figure 15 Block diagram of the Balanced Error Feedback (BEF) applied to an amplifier with nominal unity gain.

To see the details of BEF operation, we need to determine analytically the output voltage in Fig.15, which can be written as

$$\mathbf{V_o}(BEF) \;= \boldsymbol{V_s A_1}\,\frac{\mathbf{1+A_B A_2}}{\mathbf{1+A_1A_2}} + \boldsymbol{V_{e1}}\frac{\mathbf{1}}{\mathbf{1+A_1A_2}} + \boldsymbol{V_{e2}}\frac{\boldsymbol{A_1}}{\mathbf{1+A_1A_2}} \tag{30}$$

When the balance condition is met, that is when

$$\boldsymbol{A_B} = \boldsymbol{A_1} \tag{31}$$

we have

$$\frac{\mathbf{V_o}}{\boldsymbol{V_s}}(BEF) = \;\boldsymbol{A_1} \quad (\text{for } \mathbf{A_1 \cong 1}) \tag{32a}$$

$$\boldsymbol{F_{EC}}(BEF) = \frac{\mathbf{1+A_B A_2}}{\mathbf{1+A_1A_2}} = \mathbf{1} \tag{32b}$$

and the voltage at the input of the error amplifier A_2 is equal to

$$\boldsymbol{V_b - V_m} = -(\boldsymbol{V_{e2}A_1 + V_{e1}})\frac{\mathbf{1}}{\mathbf{1+A_1A_2}} \tag{33}$$

which confirms that the input signal component (**V_S**) does not appear any longer as part of the differential voltage at the input of the error amplifier A_2 (which however still needs to have **an adequately high CMRR and low common-mode induced distortion)**, and the BEF correction loop will virtually have to deal only with the non-linear error components of the output voltage **V_O**, with the following **additional benefits:**

- **relaxed requirements on the dynamic range (A_2 output voltage and current) and on the linearity of the EC circuit;**
- **intrinsically lower V_{e2} error associated with BEF error correction circuitry than AEF.**

We can also observe from eq. (30) that the residual output error V_{eo} is the same as AEF, and is given by

$$\mathbf{V_{eo}}(BEF) = \mathbf{V_{eo}}(AEF) = \boldsymbol{V}_{e1}/\rho_1 + \boldsymbol{V}_{e2}/\rho_2 \cong (\boldsymbol{V}_{e1} + \boldsymbol{V}_{e2})/\rho \qquad (34)$$

where, as usual, $\rho=(1+\mathbf{A_1 A_2})$.

If we now examine more closely eq.(32b) we can recognize the dominant and well balanced position of A_2 in both numerator and denominator, and therefore its stabilization role on their ratio. As a result we can expect relatively low sensitivity of F_{EC}(BEF) to A_B variations and hence to the balance condition. Let us consider the case where A_1, and therefore A_B, have first order transfer functions, with corner frequencies equal to f_{C1} and f_{CB}, respectively. **Fig.16** shows magnitude and phase of the F_{EC}(BEF) transfer function for a **±10% variation of the balance condition $f_{CB}=f_{C1}$**. In Fig.16 we can note that the **corresponding variations of amplitude and phase of F_{EC}(BEF) with $f_{U2}=f_{C1}$, are limited to a mere ±0.6 dB and ±3deg, respectively.** Moreover, this already low variation can be further reduced, if needed by lowering the fU2/f_{C1} ratio.

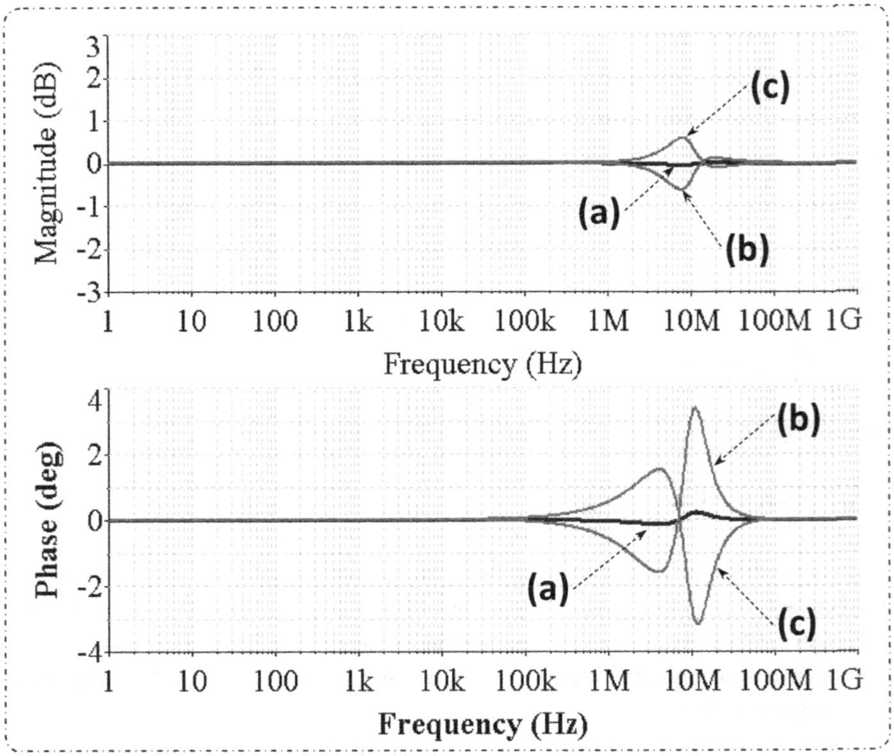

Figure 16 $F_{EC}(f)$ of BEF scheme of Fig.15 for f_{C1}=10MHz for different values of f_{CB}. (a) $f_{CB} \cong f_{C1}$ (balanced); (b) $f_{CB}=0.9f_{C1}$ (-10% unbalance); (c) $f_{CB}=1.1f_{C1}$ (+10% unbalance).

Fig. 17 shows how BEF can be practically applied to a complete audio power amplifier with gain **A_1**>1. Here there is a further block, **N**, which has the function to attenuate the output voltage V_O down to a predetermined level V_m **comparable with the balance and reference voltage V_b.**

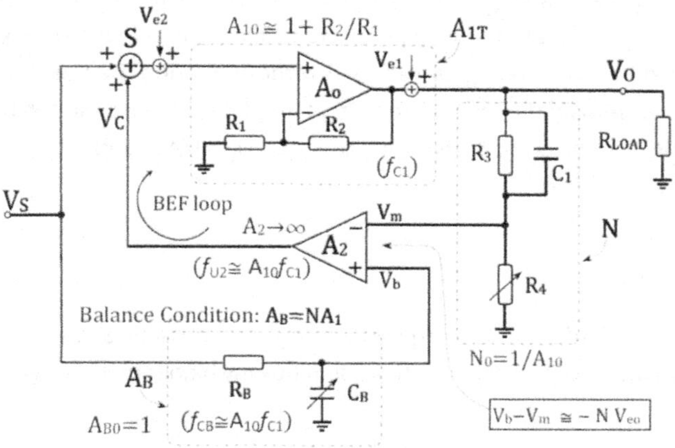

Figure 17: BEF scheme applied to an amplifier with positive gain higher than one

The output voltage is now given by

$$\mathbf{V_o} = V_s A_1 \frac{1 + A_B A_2}{1 + A_1 A_2 N} + V_e \frac{1}{1 + A_1 A_2 N} \tag{35}$$

where V_e=(V_{e1}+$V_{e2}A_1$), so stability and balance conditions for Fig.17 become, respectively,

$$1 + A_1 A_2 N \neq 0 \tag{36}$$

$$A_B = N A_1 \tag{37}$$

I can therefore write for Fig.17 the analogous equations I wrote for Fig.16, that is

$$\frac{V_o}{V_s}(BEF) = A_1 \quad (\text{for } \mathbf{A_1 \geq 1}) \tag{38a}$$

$$F_{EC}(BEF) = \frac{1 + A_B A_2}{1 + A_1 A_2 N} = 1 \tag{38b}$$

The calculations done to find a stability criterion for AEF and FDR assuming a first order transfer function for a target amplifier with unity gain, and relevant eq.(17), are still valid but must be applied not to A_1 block alone but to the block cascade **A_N= A_1N**. Therefore, eq.(17) will become

$$f_{U2} \leq f_{CN}/(4\zeta_0^{\ 2}) = f_{CN} \tag{39}$$

where f_{CN} is the corner frequency of the whole block cascade A_N, and ζ_O =0.5, as usual, is the reference damping ratio. As for f_{CN}, we can calculate it assuming a first order response for the target amplifier, that is $A_1=A_{10}/(1+s/p_1)$, where $A_{10}\cong(1+R_2/R_1)$ is the closed loop DC gain, and $p_1=2\pi f_{C1}$.

Let us now consider the role of C_1 in the circuit of Fig.17. When $C_1=0$ the corner frequency of A_N is the same as A_1, that is $f_{CN}(C_1=0)=f_{C1}$. But with $C_1\neq 0$, the transfer function of the attenuation network N will be characterized by a phase-lead pole-zero pair, p_N and z_N, with $p_N>z_N$, which can be used to extend the bandwidth of the A_N cascade, **yet without altering the transfer function of the full amplifier**, which will remain the same as the target amplifier, i.e. A_1, as stated by eq.38a. In fact if we assume $R_3/R_4=R_2/R_1$, the DC gain of block N becomes $N_O=1/A_{10}$ and the transfer function $A_N=A_1N$ can be written as

$$A_N=\frac{1+s/z_N}{(1+s/p_1)(1+s/p_N)} \tag{40}$$

and C_1 can be adjusted in order to move z_N closer to p_1, and possibly achieve $z_N\cong p_1$, although good stability margins may still be reached with z_N significantly higher than p_1. When $z_N\cong p_1$ eq.(40) becomes

$$A_N\cong\frac{1}{(1+s/p_N)} \tag{40b}$$

where $p_N=A_{10}p_1=2\pi f_{CN}$ and $f_{CN}=A_{10}f_{C1}$, which is the gain-bandwidth product of the target amplifier A_{1T}. Considering now eq.(38) which states that we can set $f_{U2}=A_{10}f_{C1}$ for the error amplifier A_2, we can realize that the introduction of C_1 with the appropriate value into the resistive network N, allows to extend the bandwidth of the BEF error correction loop up to the gain bandwidth product of the target amplifier while retaining good stability margins ($\zeta_O=0.5$). So **the error correction potential ρ_P(BEF) of the BEF scheme of Fig.17 is correspondingly increased** according to eq.(18) which becomes

$$|\rho_p(\mathrm{BEF})|\cong f_{U2}/f=A_{10}f_{C1}/f \tag{41}$$

In Fig.17 C_B (or alternatively R_B) can be adjusted, if necessary, to reach an optimum balance condition, while R_4 (or alternatively R_3) can be adjusted to fully cancel the V_s component out from the BEF error correction loop (at low frequency too!) and therefore relax the requirements on A_2 output current and voltage dynamic range. Let us also note that the BEF EC circuitry in Fig.17 does not have to manage directly the high voltages and currents of the output stage of A_{1T}, but it interacts only with the input of A_{1T}, that is with low voltage and current levels. The main consequences are that its implementation can be cheaper and more effective.

High frequency, slew-rate and clipping performance

In ordinary feedback loops (FDR), the extra gain amplifier A_2 (see Fig.7) is cascaded with the target amplifier A_{1T} and therefore has a dominant role in defining high-frequency, slew-rate and power bandwidth of the overall amplifier. AEF, HFEC and BEF error correction schemes, in which the error correction loop has only a supporting role to the main signal path (see Fig.6, Fig.10 and Fig.15), may have instead a very limited impact on these parameters. In fact, in these EC schemes, $\mathbf{V_S}$ is in theory the main driver of the target amplifier, and V_C, in the normal and almost linear operating conditions of the amplifier, only carries the small voltage contribution needed to counteract the error of the target amplifier, which is usually less, say, than 5% of the rated input signal. This is particularly true for BEF where the main signal $\mathbf{V_S}$ is virtually removed from the correction loop. A critical role in error correction schemes is played by the voltage summer where the correction voltage V_C is added up to the input voltage V_S (S4 in Fig.6 and Fig.10, and S in Fig.15). If the main characteristics of this voltage summer are well preserved in the implementation, the impact on high-frequency, slew-rate and power bandwidth of the overall amplifier can be minimized, especially in BEF, by using active devices with adequate bandwidth and slew rate (SR). A practical rule of thumb I usually apply in the selection of the Op-Amp for the implementation of the error amplifier (A_2) in Fig.17, is that $SR(A_2) \geq SR(A_{1T})/A_{10}$.

It is also very important that the input voltage V_S reappears practically intact at the output of this summer at all frequencies and levels, which entails none or very limited attenuation and very wide bandwidth for the voltage summer. The more we move away from this ideal situation, the greater the impact on dynamic performance will be. This situation changes dramatically during nonlinear (i.e. clipping and/or slew rate occurrences) and high-frequency operation (especially in AEF and HFEC) of the amplifier where V_C can become much higher than in audio-frequency linear operation and can even exceed V_S, with the risk not only to impact slew-rate and clipping performance, but also to overload one or more stages of the target amplifier (especially with high gain audio power amplifiers). However, these problems can be easily prevented by somehow limiting the corrective voltage V_C to say about 10-20% of the maximum level of the input voltage. Anyhow, let us consider that the linear operating range of the correction loop must always be enough to adequately include both the ac components of the peak error voltage to be corrected and the worst case DC component of V_C, needed to correct the offset voltage (and relevant drift) of the target amplifier.

BEF error correction iteration

In virtue of the fact that BEF EC has virtually no impact on the transfer function and dynamic performance of the amplifier it is correcting (see eq.38a and eq.38b), BEF error correction can be applied at least one more time to further reduce, if needed, the distortion of the target amplifier $\mathbf{A_{1T}}$ in Fig.15 and Fig.17. There are several ways to repeat the BEF correction process. The most direct one is to apply the second BEF correction loop to the overall corrected amplifier of Fig. 17. This can be graphically shown by replacing A_{1T} in Fig.17 with the whole corrected amplifier of the same Fig.17, as shown in **Fig.18a**, where the direct reiteration of the BEF error correction scheme is applied to our by now familiar complementary power MOSFET source follower, with its typical bandwidth of f_{C1}=10MHz.

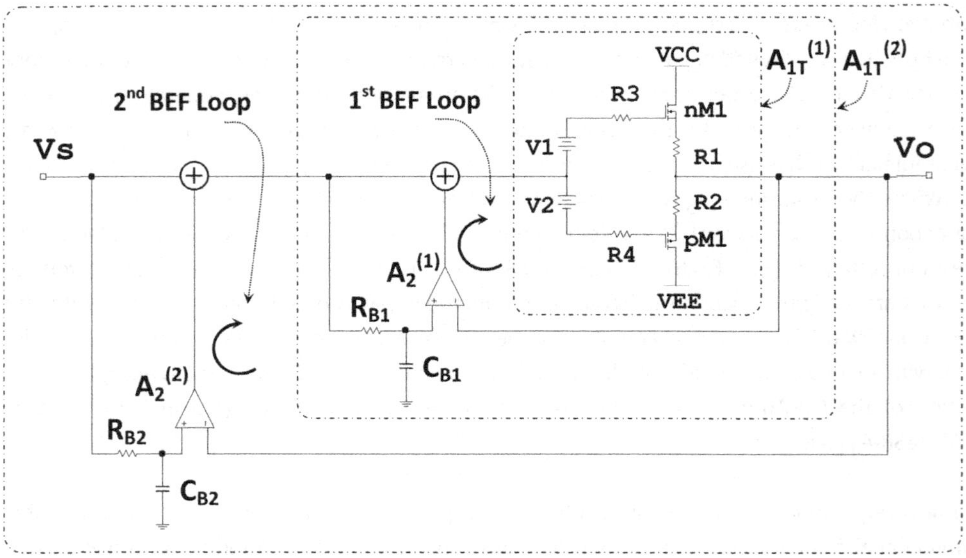

Figure 18a Direct iterative application of BEF Error Correction reduces distortion to vanishing levels.

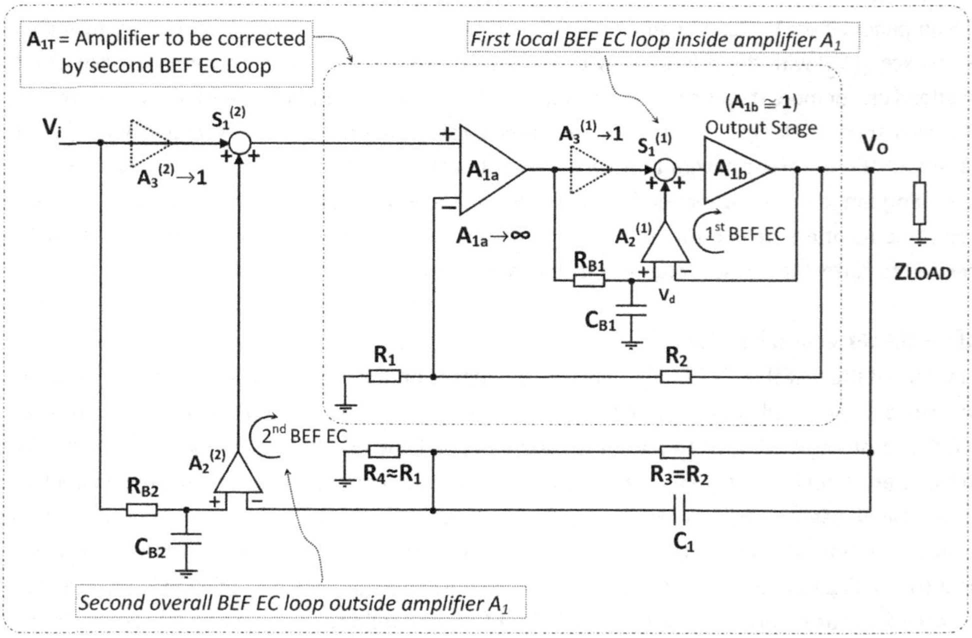

Figure 18b An example of applying iterative BEF Error Correction to reduce distortion to vanishing levels in a complete audio power amplifier.

The magnitude of the overall error cancellation ratio ρ_T in the amplifier of Fig.18a, is equal to the product of the cancellation ratios ρ_1 and ρ_2 of the two BEF EC loops (see eq.18), and in the ideal case of two identical loops is given by

$$|\rho_T| \cong |A_2^{(1)}A_2^{(2)}| \cong (f_{U2}/f)^2 \qquad (42)$$

Achieving, in concrete implementations of the system in Fig.18a, the maximum error correction potential is rather difficult, due to the increased sensitivity of the system to the balance condition. Fortunately this sensitivity can be reduced, if needed, by slightly reducing f_{U2}/f_{C1} ratio, in one or both BEF loops, without great penalties for the overall distortion reduction, which may still be very significant.

For instance, even considering a conservative A_2 bandwidth of f_{U2}= **0.3f_{C1}=3MHz**, eq.42 yields an overall error cancellation ratio of 90000 (i.e. 90dB) at f=10kHz and 900 (i.e. 59dB) at f=100kHz. So, if worst case distortion of A_{1T} is THD=0.5% (including all harmonics up to 100kHz), the residual distortion of a good implementation of the circuit shown in Fig.18a will not be worse than about 0.0005%.

Another practical way of iterating the BEF error correction scheme is given in **Fig.18b**, which shows the block diagram of a full audio power amplifier utilizing two BEF EC loops. One, inside the feedback loop of the target amplifier (trough R_1 and R_2), is employed to locally linearize its output stage, and the second one is used outside the main feedback loop to cancel the overall residual error of the whole amplifier, in particular dc offset, drift, low and high frequency noise and, of course, harmonic distortion.

This implementation is less critical because the first local BEF loop does not have stringent dc and noise requirements and its critical frequency zone (f_N) is well beyond the corner frequency of the power amplifier (f_{C1}). In general for power amplifiers with gain (A_{10}) higher than 10 (20dB), usually $f_{C1} \leq f_N / A_{10}$. This means that the impact of the internal BEF loop on the time-frequency response of the amplifier is very limited, and if needed its impact can be easily managed with properly reducing $f_{U2}^{(1)}/f_{C1}^{(1)}$ ratio.

However there is a penalty to pay: the first BEF loop has to cope with the high voltages, currents and slew-rates needed to drive the output stage, and this normally entails higher levels of circuit complexity and higher cost for the implementation (as we have seen in the above discussion). The second external BEF loop which only has to manage lower levels of voltage and current, and has lower slew rate requirements, can be implemented in a much more comfortable manner and at lower cost.

BEF review conclusions. The above BEF scheme presentation and discussion can be closed with the following summary:

- BEF error correction meets the fourth requirement for the ideal EC scheme, i.e. F_{EC}= F_{ECO}=1;
- Therefore BEF error correction has virtually zero impact on the transfer function of the target

amplifier regardless of A_2 and therefore of f_{U2}/f_{C1} ratio;

- As a result, the design of the main power amplifier (the target of the error correction process) can be focused on the optimization of its signal *transparency* and dynamic performance, while its error and distortion performance can be left in the care of the BEF loop with very limited impact on the sonic performance;
- The error correction loop and main amplifier performance can be optimized independently;
- The BEF scheme has relatively low sensitivity to the balance condition;
- BEF retains the same high level of error-correction performance as AEF;
- BEF has inherently lower requirements on dynamic range (output voltage and current of A_2) than AEF and HFEC;
- its complexity and implementation cost are low;
- it can be used inside the overall feedback loop of a power amplifier without jeopardizing dynamic stability;
- The BEF can be easily used in an iterative way to further reduce the error, if needed.

Practical applications of BEF to low distortion audio power amplifiers

I want to close this first part with the presentation of two practical applications of the BEF concept: a low distortion high speed MOSFET power voltage follower (unity gain) capable of 200W into an 8Ω load and a complete high speed low distortion audio power amplifier capable of providing 300W into a 4Ω load and 150 W into 8Ω load.

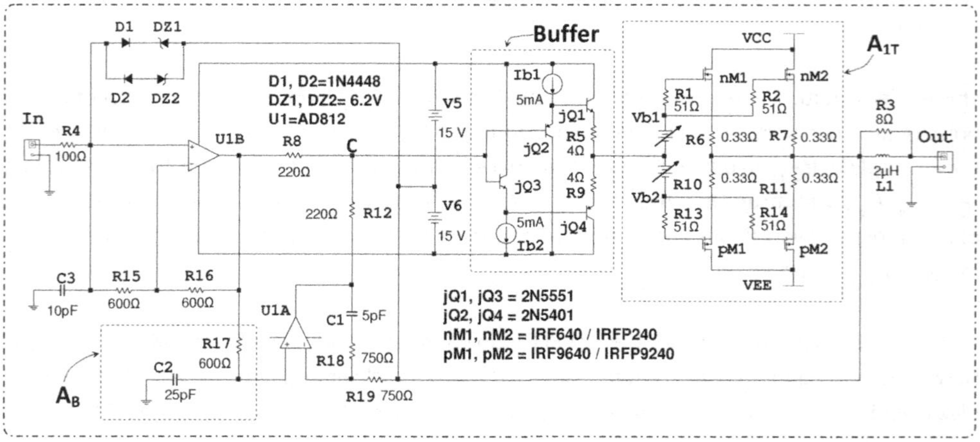

Figure 19 Low distortion 200W MOSFET Power Voltage Follower with BEF error correction (simplified diagram).

Low distortion MOSFET Power Voltage Follower (PVF). The low distortion MOSFET power voltage follower, shown in **Fig. 19** in its simplified form, employs the same high-speed dual CF Op-Amp (AD812) as its AEF counterpart of Fig. 12, but in order to increase the available output power, the output stage here features two parallel N&P power MOSFET pairs (with matched $V_{GS(TH)}$ within

10mV for both N and P devices). For the sake of simplicity, the details of the biasing network, the floating power supply and the protection circuits are not shown. In order to mitigate the problems associated with the high non-constant input MOSFET capacitance and to relax the peak output current requirement of the two CF Op-Amps of the AD812, this circuit incorporates a wide-band voltage-current buffer placed between the resistive voltage adder, made up from R_8 and R_{12}, and the output stage. This buffer (which shares the floating low-voltage dual power supply of the AD812), consisting of the four bipolar junction transistors jQ1-jQ4 in a *diamond* voltage-follower configuration, has a bandwidth larger than 100MHz and a current isolation (the ratio I_O/I_{IN} between the output and the input currents) in excess of 10^3 up to 1MHz for output currents up to 100mA. By the insertion of this separation block, the MOSFET capacitance (variations included) seen at the output of the voltage summer (node **C**) , as well as the relevant gate peak drive-current , are reduced by a factor higher than 10^3. Therefore a higher level of impedance can be set here for the resistive voltage adder network (R_8=R_{12}=220Ω), so that the AD812 can operate with more comfortable peak output currents, reduced power dissipation and, hence, increased reliability.

The balancing network A_B is simply implemented with an RC low-pass filter, R17 and C2, with a corner frequency f_{CB} of about 10MHz. Unity gain frequency $f_U(A_2)$ of the error amplifier (built around Op-Amp U1A) is set to about 5-10MHz by means of the feedback network C_1≅5pF and R_{18}=750Ω, in order to maintain good stability margins in the whole operating range of the PVF. Components R_4, D_1, D_2, DZ_1 and DZ_2 have been introduced to protect the circuit against input overload. Their task is to prevent that the input- output voltage difference exceeds a safe limit (of about 7V, here). When this occurs, the excess current which is forced to flow through R4, is simply shunted to the output. The side effect of this input voltage limitation (which also limits the maximum output current of the input stages of CF Op-Amps U1A –U1B and hence their slew rates) is that the slew rate of the overall PVF is limited to about 200-300V/μs which, however, is more than enough for all possible applications in audio power amplification.

Low distortion PVF performances. Accuracy and distortion performance of this PVF (powered by ±65V) are very good and similar to those of Fig.12; that is:

- Offset voltage less than 10mV;
- THD at V_{out}= 56.5V_P into a load ZL=8Ω (i.e. about 200W) is less than 0.0002% from 20Hz to 2kHz, and lower than 0.003% at 20kHz;
- Output swing at clipping: +63V to -62V.

I am planning to use the PVF in Fig.19 as the output stage of a very low distortion audio power amplifier with two BEF error correction loops (in line with Fig.18b) which is currently in the design phase. The ambition is to achieve THD performance below 0.0005% at all frequencies and power levels.

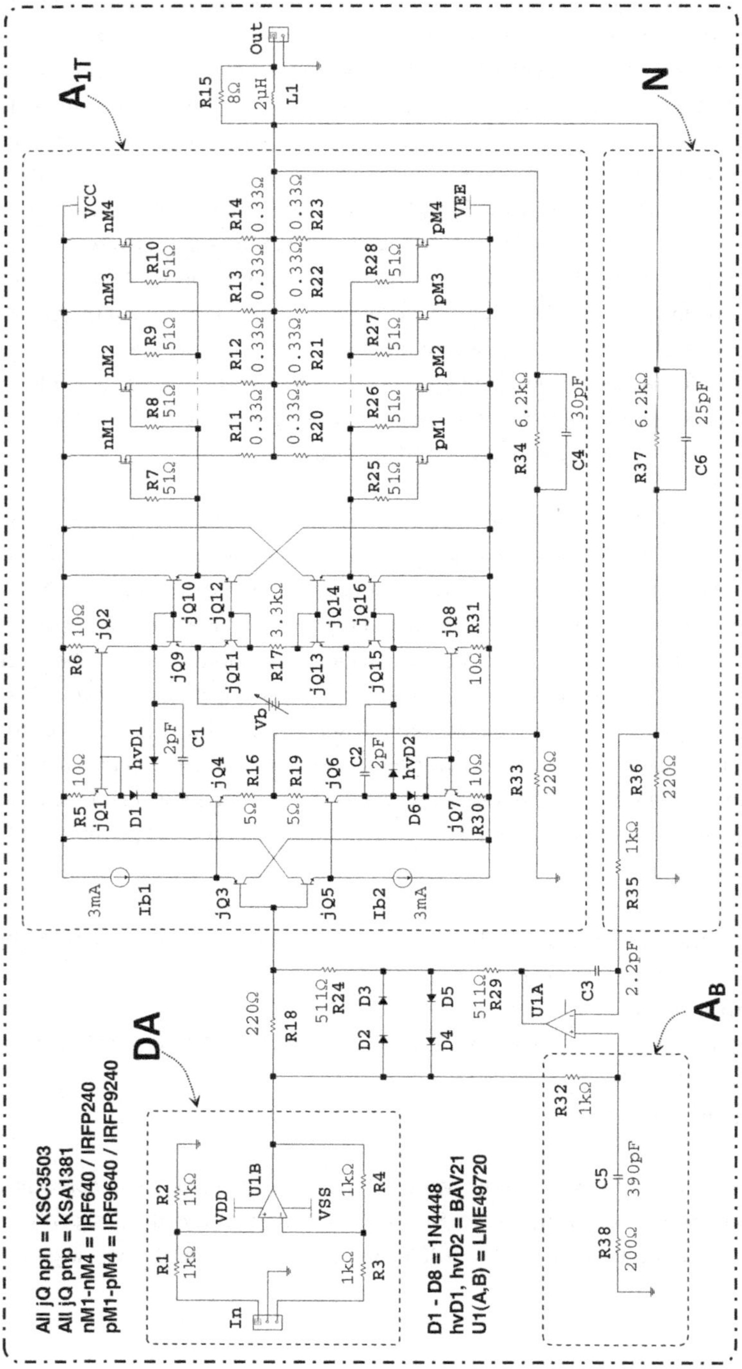

Figure 20 High Speed Low distortion 200W Audio Amplifier with BEF error correction (simplified diagram).

High-speed low-distortion audio power amplifier. The second example I want to present and discuss now is the application of the BEF EC scheme of Fig.17 to the implementation of the low distortion high-speed audio power amplifier shown, in its simplified form, in **Fig.20** (details of biasing and protection circuits are not shown).
The target amplifier for the error correction, here, is a power amplifier with a gain of about 29dB which employs a current feedback (CF) architecture. The main typical advantages of CF architectures compared to conventional voltage feedback architectures, are well-known:

- single voltage-gain stage which in general results in wider open loop bandwidth and better high frequency behavior (virtually single pole open-loop frequency response) and associated lower linear distortion;
- all stages are push-pull class AB operated, with a high output peak- current/idle-current ratio (yet the maximum current of the bipolar devices is cautiously limited to 15-20mA to avoid excessive stress) ;
- large signal response very similar to small signal response (both in the time and in the frequency domain);
- high slew rates and absence of slewing phenomenon in well-designed CF audio amplifiers;
- high signal transparency (for all the above reasons).

The main disadvantages of CF architectures are also well-known:

- lower dc gain;
- higher offset voltage and drift;
- higher output noise;
- higher distortion in the audio frequency range.

Fortunately, these disadvantages can be well counteracted by BEF (as well as by AEF) error correction, so very good performances may be achieved even by the application of a single BEF loop.

Circuit description. The target amplifier, A_{1T}, consists of a three-stage amplifier with its own feedback loop, R_{33} and R_{34}, with closed loop gain equal to $\mathbf{A_{10}=1+R_{34}/R_{33}}$ (about 29dB). The input stage is a push-pull class AB voltage to current converter, whose transconductance $\mathbf{g_m\cong 1/R_{33}}$ is determined by resistor R_{33}, which is also part of the feedback network. Input stage transistors jQ3-jQ4-jQ5-jQ6 form a *diamond* voltage-follower with the task to produce the copy of the input voltage to be applied to resistor R_{33} in order to generate the current $\mathbf{I_1=V_{i1}/R_{33}}$ which feeds the intermediate stage. The intermediate stage is made up from two complementary Widlar bipolar current mirrors, jQ1-jQ2 (pnp) and jQ7-jQ8 (npn), features a full output voltage swing (from $+V_{CC}$ to $-V_{EE}$), and has the task to produce a copy of the input current, which in turn is used to drive the input impedance of the output stage. The output stage consists of the buffer made up from transistors jQ9 to jQ16 and up to four parallel N&P power MOSFET pairs (nM1-nM4 & pM1-pM4).

Both N and P devices have been selected for $V_{GS(TH)}$ within 10mV. Current sources $\mathbf{I_{b1}}$ and $\mathbf{I_{b2}}$ set the bias condition of all bipolar transistors, while the DC voltage $\mathbf{V_b}$ is adjusted to set the idle current

in the MOSFET devices to about 100mA each, which is enough to ensure for the output stage a bandwidth of about 10 MHz (for $\mathbf{V_{out}= 0}$) and an open loop output resistance as low as 0.05Ω.
A precision, low noise and low distortion conventional voltage-feedback dual Op-Amp IC (LME49720), with an adequate gain-bandwidth (GB) product (typically GB=55MHz) and slew-rate (typically±20V/μs), is used to implement the BEF error correction circuit. One Op-Amp (U1B) is used to implement the low-noise input differential amplifier (DA) of this power module (designed to be connected to the input signal front-end located on a different board of the audio equipment). The other Op-Amp U1A is used as the error amplifier A_2 of the BEF loop. The unity gain frequency f_{U2} (≤GB) of the error- chain, equal to $A_2\,[R_{18}/(R_{24}+R_{29})]$, is set to the appropriate value (between 5 and 10 MHz) by means of the feedback capacitor C_3. The stability margins of the BEF loop are adjusted by means of the phase-lead capacitor C_6 of block **N**, which moves $\mathbf{z_N}$ of eq. (40) towards $\mathbf{p_1}$ and has also a role in the optimization of the high-frequency distortion of the amplifier (to partially compensate for some layout and pcb design imperfections and limited common mode performance of the used Op-Amps). The balancing block A_B is simply implemented with an RC network (formed by the two resistors R_{32} and R_{38} and the capacitor C_5) which can emulate almost perfectly the frequency response of the block cascade $A_{1T}N$. The voltage summer S in Fig.17, is implemented with a resistor network, R_{18}, R_{24} and R_{29}, which has an attenuation of about 1.7 dB along the main signal path and of about 14 dB along the error-correction path. Diode clamping network D_2-D_3-D_4-D_5 has the function to limit the correction voltage appearing across R_{18}, for the reasons discussed above, to a safe ±350mV, which represents a ±20% of rated peak input voltage, so that the risk of input overload due to the EC circuit is avoided.

Let us note, incidentally, **compactness**, **simplicity** and **low cost** of the BEF circuitry, which is not more than 2-3 Euros.

Amplifier performance

- **Supply voltages.** Nominal supply voltage range is: $\mathbf{V_{CC}=-V_{EE}= 60 \pm 5\ V}$. All following measurements have been made with V_{CC} =55V and V_{EE}=-55V (from two 0-60V/0-3A adjustable voltage regulators).;
- **Output power.** The amplifier of Fig.20 can safely provide 100W into an 8Ω load and 200W into a 4Ω load (150W into an 8Ω load and 300W into a 4Ω load with V_{CC} =65V and V_{EE}=-65V)
- **Time-Frequency Response.** Frequency and square-wave response are shown in **Fig.21** and **Fig.22**, respectively (simulation results are in perfect agreement with measurement results). The two figures demonstrate that the time-frequency response of the overall amplifier is virtually left unaffected by the application of the BEF error correction loop. Measured **gain** and **bandwidth** (-3dB) are 29dB and 330 kHz, respectively, with and without BEF EC;
- **Slew-rate.** With four N&P MOSFET pairs, measured maximum slew rate with rated input voltage, V_S=2V_P, is ±100V/μs (with a 10kHz square wave input signal, with and without BEF EC). **Note:** slew-induced distortion is virtually absent since during transients all stages of the amplifier continue to operate in their linear range, even if maximum output currents of the input and inter-

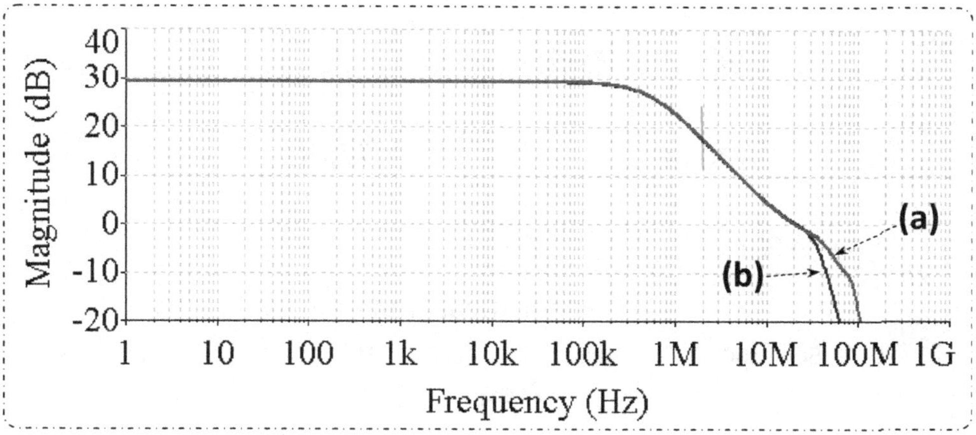

(a) BEF error correction not-active

(b) BEF error correction active with optimal setting of balancing block A_B

Figure 21 Frequency response of the High Speed Low Distortion 200W BEF Audio Power Amplifier of Fig.20.

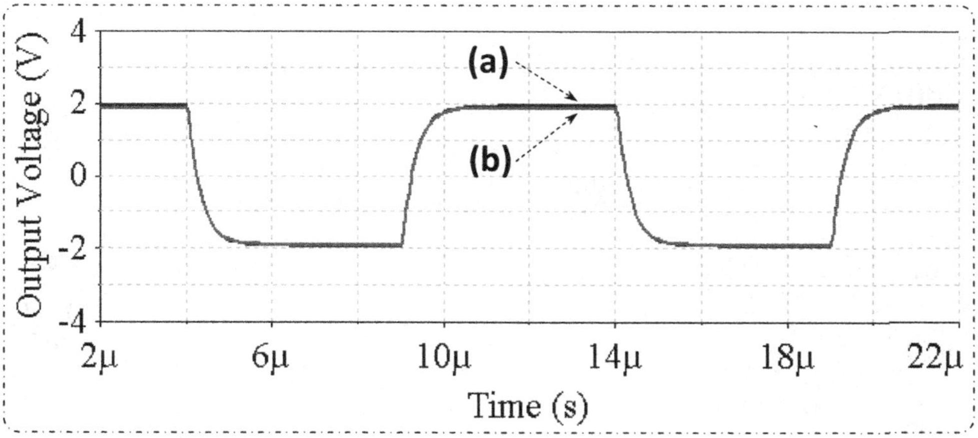

(a) BEF error correction not-active (with 1.3V offset removed to simplify comparison);

(b) BEF error correction active with optimal setting of balancing block A_B

Figure 22 100kHz square-wave response of the High Speed Low distortion 200W BEF Audio Power Amplifier of Figure 20 with V_O=4V_{P-P}.

mediate stages are limited to about 15mA;

- **Power Bandwidth.** This amplifier has been designed to safely provide 100V_{P-P} continuously into 4-8Ω for frequencies up to 100kHz. However, it can occasionally sustain a voltage swing of 100V_{P-P} up to 300 kHz for short periods of time (cautiously say <300ms). The reason for time limitation is the fact that at frequencies *f*>100 kHz the power dissipation of the bipolar transistors can exceed the safe operation limits;

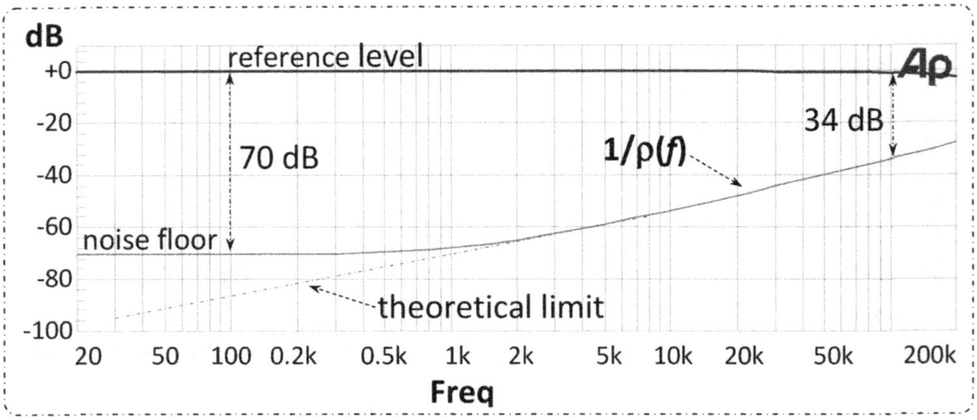

Figure 23 Measurement results of the inverse of the error cancellation ratio 1/ρ(f) achieved by the BEF error correction loop in Figure 20. Test conditions: V_e=1V (0dB); MBW >500 kHz.

- **Error cancellation ratio. Fig.23** shows the measured inverse of error cancellation ratio versus frequency, which is perfectly in line with theoretical predictions and simulations: higher than 70dB up to 1 kHz (limited by noise) and about 34 dB at 100 kHz.

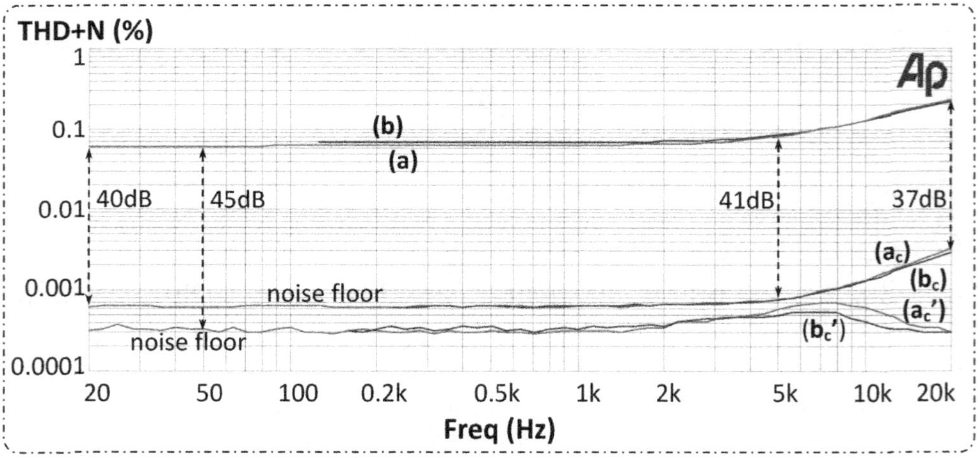

Measurement Bandwidth (MBW)=22Hz-80kHz:

(a) Amplifier with no BEF error correction, Load Impedance=∞

(b) Amplifier with no BEF error correction, Load Impedance=8Ω (P_{OUT}=100W)

(a_C) Amplifier with BEF error correction, Load Impedance=∞

(b_C) Amplifier with BEF error correction, Load Impedance=8Ω (P_{OUT}=100W)

Measurement Bandwidth (MBW)=22Hz-22kHz:

(a_C') Amplifier with BEF error correction, Load Impedance=∞

(b_C') Amplifier with BEF error correction, Load Impedance=8Ω (P_{OUT}=100W)

Figure 24a THD+N vs frequency of the High Speed Low Distortion BEF Audio Power Amplifier of Fig.20. Test equipment: Audio Precision System Two. Test Conditions: Val=±55V, V_O=80V_{P-P}

- **Offset Voltage.** The output offset voltage of the target amplifier is drastically reduced by the BEF loop from 1.3V to about 25mV;
- **Output noise.** The output noise (bandwidth: 22Hz-22kHz) is reduced by the BEF loop from 150μV to about 65μV;
- **Total Harmonic Distortion.** Measurement results of THD+N (%) versus frequency for P_{OUT}=100W into 8Ω are shown in **Fig.24a.**

The distortion of the target amplifier before correction is 0.06% at low frequencies and increases up to 0.23% at 20kHz (MBW=80kHz). With the activation of the BEF error correction loop, **the distortion is virtually removed at frequencies below 1kHz**, where the THD+N residue (0.0006% for MBW≅80 kHz, and 0.0003% for MBW≅22 kHz) is dominated by noise, and is drastically reduced by about 40 dB (i.e. a hundred times) for frequencies ranging from 1kHz to 20kHz, where THD+N (MBW≅80 kHz) is a mere 0.003%, in perfect agreement with simulation results. At lower output power levels this high value of distortion cancellation ratio is retained and THD+N (%) tends to be dominated by noise at all audio frequencies.

BEF capability to act effectively against distortion also in conventional voltage-feedback amplifier architectures, is demonstrated by **Fig.24b,** which reports the measured THD+N (%) performance of a conventional (low THD) audio power amplifier with BEF error correction. Schematic and component values are the same as Fig.20, with the only exception of the target amplifier A_{1T}, which retains the

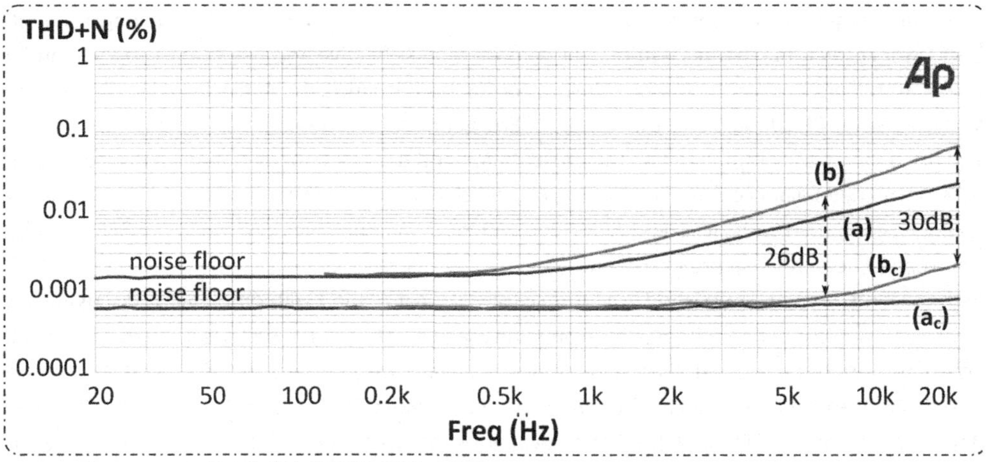

Measurement Bandwidth (MBW)=22Hz-80kHz:
(a) Amplifier with no BEF error correction, Load Impedance=∞
(b) Amplifier with no BEF error correction, Load Impedance=8Ω (P_{OUT}=100W)
(a_C) Amplifier with BEF error correction, Load Impedance=∞
(b_C) Amplifier with BEF error correction, Load Impedance=8Ω (P_{OUT}=100W)

Figure 24b THD+N vs frequency of a conventional Low Distortion Audio Power Amplifier with BEF Error Correction. Test equipment: Audio Precision System Two. Test Conditions: Val=±55V, V_O=80VP-P on8Ω.

same feedback network and output stage (with relevant buffer and biasing circuit) shown in Fig.20, as well as the same low frequency gain (A_{10}) and corner frequency (f_{C1}), but incorporates more conventional input and intermediate stages, operating in class A. Fig.24b shows that the BEF EC loop reduces distortion by about 30dB from 5 kHz to 20kHz, where THD+N reaches its highest, 0.002%. Below 5 kHz, THD+N (≅0.0006%) is virtually noise only.

Clipping. The amplifier has clean clipping performance and corresponding low recovery times, as shown in **Fig. 25:**

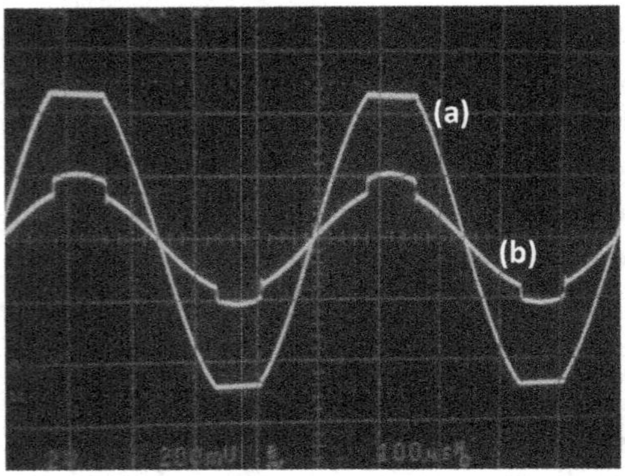

Figure25 Clipping behavior of the amplifier in Fig.20 with V_{CC}=+55V, V_{EE}=-55V, Load=8Ω, f=2kHz

***(a)** Output voltage (20V/div)*
***(b)** Input voltage of the target amplifier with correction voltage **V_c** clamped to ±350mV (2V/div)*

Summary

In this first part of my contribution I presented the results of my investigations on two popular feedback error correction schemes (**FBEC**), which I have found effective and practical to apply to audio power amplifiers, in order to improve their distortion performance: the Active Error Feedback (AEF) scheme by J. R. Macdonald and the Feedback Error Correction scheme by M. J. Hawksford (HFEC). The discussion was completed with the presentation and the assessment of a new error correction scheme, named Balanced Error Feedback (BEF), which features some distinctive and important advantages compared with HFEC and AEF. A few practical and effective applications of BEF to audio power amplifiers were eventually presented.

In the second part to be published in a future issue of Linear Audio, I will complete my discussion on error-distortion correction schemes, by examining the two last ones, the first based on ***true*** (i.e. in full accordance with the original Black scheme) feed-forward error correction **(FFEC)** and the second based on mixed feed-back and feed-forward error correction **(FB-FFEC)**. I will also include the latest developments of the FFEC implementation technique that was presented in 1994 **[13]**. These developments have been employed more recently (2007) in the 200W audio power amplifier of the *superLinear* demo-system (Pre-Amplifier and Power Amplifier), featuring a total harmonic distortion lower than 0.0005% and an intermodulation distortion less than 0.0001%. More specifically I will focus on the improvements that were introduced:

- in the output voltage summer, developed around a small ferrite wideband transformer and incorporating a main signal flux cancellation technique, in order to improve the error cancellation

ratio and extend the useful bandwidth of the error correction loop;
- in the auxiliary amplifier architecture, in order to simplify the calibration procedure, to avoid overload of the output transformer, and to achieve a better distortion performance of the overall corrected amplifier.

I will complete my contribution with the presentation and discussion of a new hybrid scheme employing in a synergetic way both **BEF** and **FFEC** error correction schemes to potentially achieve in audio power amplifiers, harmonic and intermodulation distortion levels stably lower than 1ppm (0.0001%) at all audio frequencies and power levels.

References

[1] M. J. Hawksford, "*Distortion Correction in Audio power Amplifiers*", J. Audio Eng. Society, Vol.29, No.1/2, 1981 Jan/Feb, pp. 27-30

[2] M. J. Hawksford, "*Power Amplifier Output Stage Design Incorporating Error Feedback Correction With Damping Enhancement*" Presented at the AES 74th Convention 1983, October 8-12, New York

[3] H. S. Black, "*Translating System*", U.S. Patent 1686792 (1928 Oct. 9)

[4] H. S. Black, "*Wave Translation System*", U.S. Patent 2102671 (1937 Dec. 21)

[5] W. Baggally, "*Distortion Cancellation in Audio Amplifiers*", Wireless Eng. & Experim. Wireless, Vol.10, pp.413-419 (1933 Aug)

[6] J.R. Macdonald, "*Active Error Feedback and its application to a specific driver circuit*", Proceeding of the IRE, July 1955,pp. 808-813

[7] R. J. Cordell, "*A MOSFET Power Amplifier with Error Correction*", J. Audio Eng. Society, Vol.32, No. 1/2, 1984 Jan/Feb, pp. 2-17

[8] J. Didden, "*paX-a Power Amplifier with Error Correction*", Elektor N.4, 2008, pp.24-32

[9] B. H. Candy, "*Wideband Op. Amp. Having a Plurality of Feedback Loops*", U.S. Patent N. 6052027 (2000, Apr. 18)

[10] E. M. Cherry, *"A new Result in Negative Feed-back Theory, and Its Application to Audio Power Amplifiers"*, Int. J. Circuit Theory, vol. 6, pp. 265-288 (1978 July)

[11] E. M. Cherry, "Nested *Differentiating Feedback Loops in Simple Audio Power Amplifiers*", J. Audio Eng. Society, Vol.30, No. 5, 1982 May, pp. 295-305

[12] G. Stochino, A. Pantaleoni, "*Amplifier Device With Reiterable Error Correction Scheme With Balanced Negative Feedback*", Pub. No. U.S. 2012/0326790 A1 (2012, Dec. 27)

[13] G. Stochino, "*Audio Design Leaps Forward?*", Electronics World + Wireless World, Vol. 100, No. 1703, Oct. 1994, pp.818-827

The Equal Opportunity: A Balanced Moving Magnet Phono Stage

Stuart Yaniger

"All progress is precarious, and the solution of one problem brings us face-to-face with another problem."
–Dr. Martin Luther King, Jr.

Introduction

The popularity of vinyl LP record sources in the high end home audio segment continues unabated, despite the overwhelming technical superiority of digital recording and playback. Vinyl aficionados tout a greater degree of "naturalness" and "musicality," while the detractors point out the poor signal to noise ratio, high distortion, poor separation and crosstalk, and inferior pitch stability. Without diving into that heated verbal battle, my strong feeling is that if one wishes to use an anachronistic technology, one should at least understand the engineering issues and bring the best technological solutions to bear.

One of the few places where the electronics part of the signal path has high criticality is the phono preamplifier. Low noise, accurate frequency response, high overload margins, and rapid recovery from overloads can be challenging for the designer. I will briefly consider each of these factors in turn and see how they impact on design and help us focus on design requirements. And in a moment of foreshadowing, the design topology that is used here is suitable as a gain block for many other applications, such as input stages for balanced or push-pull amplifiers.

Motivation

Although any "serious" audiophile sneers at anything other than a low output MC, preferably handmade by a reclusive Japanese craftsman and selling for 5 figures, there remains a cadre of vinyl lovers who swear by moving magnets, claiming that they have a sound closer to that of master tapes. There are certainly excellent cartridges using both technologies, but I personally have not been able to make that comparison in a controlled manner. That confessed, the two cartridges whose playback reminded me the most of master tapes happened to be MMs, despite the fussier and more critical setup that they require compared to MCs. Overcoming my intrinsic laziness, I decided to plunge back in and see what today's MMs are like. After poking around, one good candidate for Best MM seemed to be the Audio-Technica 150MLX. There were many things that appealed- boron cantilever, line contact stylus, high trackability. But there seemed to be strong divides of opinion among Internet posters, with some claiming a flat and neutral tonality and others complaining about excessive brightness and glare. And I wondered about the setups of these users, including most especially the preamp.

Moving magnet and moving coil cartridges act very differently and have different design requirements. MCs have low output (0.2mV or less is common), low source resistance (on the order of 10R), and a relatively flat impedance over the audio band (20Hz – 20kHz). MMs, by contrast, have higher output (5mV typical), moderate source resistance (on the order of 1k), and quite high inductance (on the order of 1H), and therefore have a source impedance that is anything but constant across the audio band. Both major types of cartridge are velocity sensors, that is, their output is proportional to stylus velocity and thus frequency for a constant amplitude signal. This is the basis for the required RIAA equalization curve.

Because the MC and MM sources are electrically so different, a preamp design optimized for one is unlikely to be optimized for the other. Even if one takes into account the level differences and adjusts gain accordingly, the difference in source impedances is highly significant and impacts both noise and frequency response.

The one thing that each of these types of cartridge has in common is that they are nearly always intrinsically balanced sources[1]. Interestingly, despite the low output of phono cartridges, MM or MC, the consumer audio industry almost exclusively interfaces them to their preamps using an unbalanced connection, contrary to standard engineering practice in any other metrology application. To address this curious oversight for MC users, I developed a balanced input preamp suitable for their low output voltage and source impedance.[2] However, the techniques used are not suitable for MM cartridges, so to fill that niche, I hereby present The Equal Opportunity Preamplifier. In Part 1, I will examine the engineering issues and develop the circuit. In Part 2, I will present a practical implementation with associated measurements. Our kind Editor has graciously agreed to design PCBs for ease of construction.

Design Target

What we ideally would want from an MM preamp is:

- Low noise
- Conformance to RIAA when interfaced to expected inputs and outputs
- Sufficient gain to match modern sources
- Low distortion
- Low source impedance and the ability to drive any reasonable length of interconnect
- Balanced input and output

[1] There are a few exceptions, most notoriously the Decca cartridges. Some other cartridges (e.g., Shure), come with one pin electrically tied to the case, but that strap is external and easily cut away.

[2] Stuart Yaniger, "His Master's Noise" http://www.diyaudio.com/forums/diyaudio-com-articles/163570-his-masters-noise-thoroughly-modern-tube-phono-preamp.html

- Overload margins high enough to never be exceeded, and/or
- Instantaneous recovery after overload (no blocking, sticking to rails, oscillation)

Additional Constraints

Some aspects of design are esthetic choices, rather than functional. As long as they don't significantly impact audible performance, they can be indulged. And thus my one esthetic constraint- I want to use vacuum tubes for all voltage amplification tasks. Although this choice gives up a sliver of ultimate noise performance and increases cost, it does bring along higher overload margins and greater intrinsic linearity (or potentially so) device-for-device. Solid state circuitry will freely be used in this design for achieving constant voltages, constant currents, and buffering; for these functions, the performance, complexity, and reliability of tubes is demonstrably inferior. The Equal Opportunity Preamp offends both tube purists and ultra-performance solid state designers equally.

Overview of the Design Issues

Issue 1: Noise

The principle job of a phono preamplifier is to not make noise. To put some hair around that statement, we want the noise to be low enough to be inaudible at any reasonable volume setting. There are three types of noise that are significant here.

First, the input noise of the cartridge itself. A MM cartridge can be modeled as a resistance in series with an inductance. The resistance has an intrinsic thermal noise (Johnson noise) given by the equation $V_n = \sqrt{(4KTR\Delta f)}$, with K being the Boltzmann constant, T the absolute temperature (Kelvin), R the DC resistance (DCR), and Δf the bandwidth of the measurement.

Here is where things get a bit more complicated than for MCs. MMs are designed with very specific loading requirements, generally including an optimal resistance and capacitance in parallel. Nearly everyone has settled on 47k for a resistance; optimal capacitance can be anywhere from 100-700pF, depending on the cartridge (about which, more later). This resistance, of course, has its own Johnson noise. MCs, with their low DCR and inductance, largely shunt this noise to ground. For MMs, at low frequencies where the inductive reactance of the cartridge is low, the noise is shunted less effectively due to the higher DCR. The shunting reduces further at high frequencies due to the cartridge's inductance. The load capacitance for the cartridge (i.e., the input capacitance of the preamp) has an increasing shunting effect at high frequencies, but of course resonates with the cartridge inductance.

Another inherent input noise source from the cartridge is the current noise of the cartridge's DCR dropped across the cartridge's inductance. This noise is correlated with the cartridge's voltage noise, but has rising amplitude with frequency.

Evidently, the calculation of the noise inherent to the source, before the amplifier stages have a chance to touch the input signal, is not straightforward. Yet this is the number which sets a lower bound to system noise performance- a perfectly noiseless preamp can't do any better than the noise from the source. So we need to quantify this to set a rational target for preamp noise.

Fortunately, about 35 years ago, National Semiconductor published an applications note which performed all of the mathematical heavy lifting[3]. Their derived equations can easily be programmed into an Excel spreadsheet (one of the great blessings of the 21st century), the basic cartridge and load parameters entered, and noise estimated. I have posted the appropriate spreadsheet on my website[4]- it can be quite interesting to see how loading and cartridge parameters affect noise and the results conform remarkably often with the rule of thumb offered by Tim deParavicini that phono systems are limited to 68dB signal-to-noise, no matter what the preamp design. For the Audio-Technica 150MLX that I use, the best-case signal-to-noise ratio is about 65dB.

Second, we have the noise caused by the preamp itself. This has been beaten to death by many authors, so I won't deign to repeat their work. Suffice to say that the important parameters are the equivalent input voltage noise density (e_n) and the input current noise density (i_n). The latter becomes increasingly important with higher source impedances- including reactive ones- by transforming to voltage noise via the source impedance. For FET and vacuum tube circuits, i_n is essentially zero; preamps with bipolar transistors at the input can suffer from current noise effects, which often swamp the voltage noise if the source impedance is sufficiently high. That's why bipolar transistors are often a good choice for MC preamps, but FETs and tubes can be preferred for MM.

Third, we have noise picked up by the cartridge and the wires connecting it to the preamp. The reality is that we are bathed in all sorts of electrical noise, and if you have a low output transducer with big coil and connect it with a hunk of wire to the input of a high gain preamp, noise is inevitable. So ideally, we would run our cartridge balanced, have a balanced input, and a preamp with high common mode rejection (CMR), since our likely noise pickup will assuredly be common mode. The fussiness of hum reduction using single-ended preamp inputs for cartridges is legendary, and yet totally unnecessary.

Issue 2: Frequency Response

Generally, in the context of phono replay, "frequency response" is meant as a proxy for "conformance to RIAA equalization standards." This is a necessary but not sufficient condition for accurate preamp performance, so we will consider it first.

The most common way of implementing the RIAA equalization curve is based on a flat (or nearly so) frequency response gain block, with a feedback loop containing the equalization time constants

[3] National Semiconductor, "Audio Radio Handbook," 1980
[4] http://syclotron.com/?p=121

(3180us, 318us, 75us) wrapped around the gain block (see **Figure 1** for the basic topologies). This can be an excellent solution or a terrible one. First, this method needs a lot of open loop gain- the RIAA characteristic spans nearly 40dB over the audible range (20Hz-20kHz), and the signals must be further amplified to bring them from millivolt to volt level. So in order to get good conformance to the RIAA equalization curve, the open loop gain needs to be considerably higher than that, at least 80dB. For IC opamps, that's relatively easy, since the gain reduction past the dominant pole more or less follows the same 6dB/octave slope as the RIAA curve, and the gain is highest at low frequencies, where RIAA gain also is the highest. By contrast, tube and discrete FET circuits are a bit more problematic. Their limited open-loop gain can cause significant deviations in the equalization and distortion, especially at lower frequencies. Some designers (notably Dynaco) added in positive feedback to get the open loop gain high enough within material constraints, but this is done at the expense of linearity and LF stability.

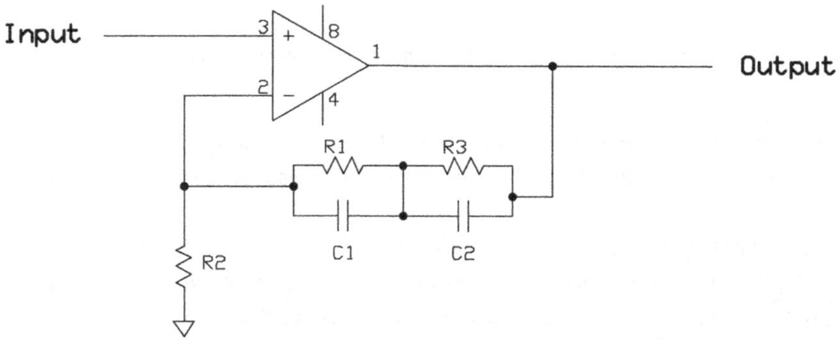

Figure 1 Conventional feedback RIAA preamp topology.

Often overlooked is the need for the output of the phono preamp's gain block to drive the feedback loop at high frequencies. Returning to Figure 1, we can see that the feedback network has a very low impedance at high frequencies, increasing the loading of the preamp output. So we'd want R2 to be as large as possible (at the expense of noise) or build in exceptional drive characteristics to the preamp's output. Many "classic" tube preamps fail right there, showing increasing high frequency distortion and gross overload when the cartridge mistracks.

Another popular topology for implementing the RIAA equalization curve is shown in **Figure 2**. A passive circuit, usually comprising resistors and capacitors (but in certain fashionable circles, using inductors and resistors), is sandwiched between two gain stages. The output is then buffered, except in the cheapest and lowest performance designs. The RC circuit calculations have been laid out in exquisite detail and clarity by Lipshitz[5]. Sometimes, the time constants are split, with the 3180us and 318us time constants placed between the first and second gain stage and the 75us time constant placed between the second gain stage and the output buffer. The choice depends on signal levels, overload margins, and

[5] Stanley Lipshitz, "On RIAA Equalization networks," JAES 27:6, 458, 1979.

noise, though I have a sneaking suspicion that the split time constant is chosen for calculational convenience. Nonetheless, Morgan Jones gives a rule of thumb[6] that if the first stage outputs more than 200 mV, it's best to put all the time constants together in the sandwich, and if it's less, it's better to split them. There are no explicit calculations to show where that number comes from.

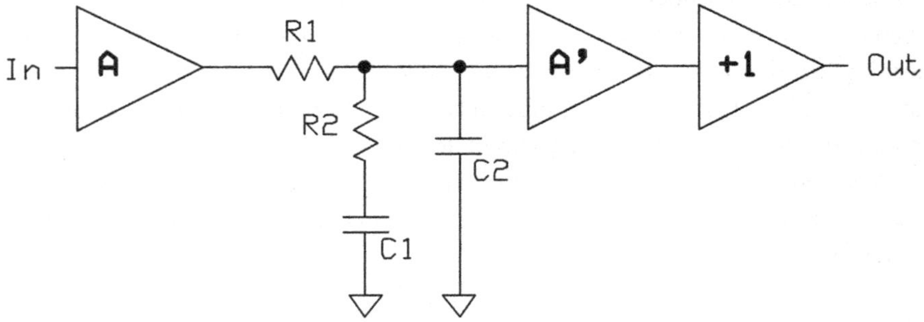

Figure 2 Passive RIAA equalization, sandwiched between two gain stages.

The principle advantages of this second approach include a high tolerance for gain variations, good overload margins, and low distortion. With a bit of care, the noise can be made low as well.

Once we have correctly implemented RIAA within the preamp, the most overlooked issue is effects of preamp input impedance on frequency response. This is due to the high DC resistance and inductance of MM cartridges. As mentioned above, all preamps for MM have a recommended input resistance and capacitance. These values are generally specified by the cartridge manufacturer and interact with the mechanical properties of the cartridge to provide an *overall* flat frequency response. If one deviates from these recommendations, the response will likewise deviate. For some, this is a useful tool for "tuning" the sound of a system, in a sense an opportunity for equalization. But it's just as likely, if not more so, that this uncontrolled variable will swamp any of the other frequency response errors in the system.

Let's look at an example of this in practice. The Audio-Technica 150MLX that motivated this article has a DC resistance of about 2k and an inductance of 0.45H. The recommended load for it is 100-200pF. Now, in practice, what is it likely to see? First, there's the capacitance of the wires running from cartridge to tonearm output plug. Those measure about 15pF for my tonearm. Then there's the cable running from the tonearm to the preamp. When researching this article, I was stunned that this cable capacitance is rarely specified, even when people are paying ridiculous premiums for audiophile fashion statements. Measuring a few samples I had laying around, 150pF seems like a pretty typical number for a meter-long cable. This could of course be reduced by building the preamp right into the turntable, but that brings in a lot of other design issues. We're up to 165pF before the preamp is in the picture.

[6] Morgan Jones, "Valve Amplifiers," 5th edition, Newnes, 2012.

A "classic" tube phono stage (e.g., Marantz, Dynaco, Audio Research) will use a 12AX7 in the first stage. The gain using their parts values is roughly 70. 12AX7 grid-to-plate capacitance is specified at 1.7pF, and usually runs a bit higher than that; several tested samples of 12AX7/ECC83 for various manufacturers had figures closer to 2.5pF. Wiring and strays add another 4-5pF to that. It's then readily determined that the input capacitance from the Miller effect (which swamps the grid-to-cathode capacitance) is close to 500pF! Even if we select tubes for low capacitance and wire carefully, the best we could hope for is perhaps 350pF. FET circuits will show similar results, with a larger contribution from the gate-to-source capacitance. Let's take the Panglossian figure of 350pF, add the cable capacitance and go with that, for a total load of 500pF on the cartridge. What is the effect on the frequency response?

Start by assuming that the engineers at Audio-Technica know what they're doing and that the flattest frequency response comes from loading the cartridge according to their recommendation. If we use that electrical response as the reference and assume that mechanical response is unchanged with loading (a reasonable assumption), we can see how the overall frequency response changes when the capacitance of the load is increased to 500pF. This is plotted in **Figure 3**; note that the response rises in the treble, with a 3.3dB peak at 10 kHz. Is it any wonder that while some users love this cartridge, others complain that it's too bright?

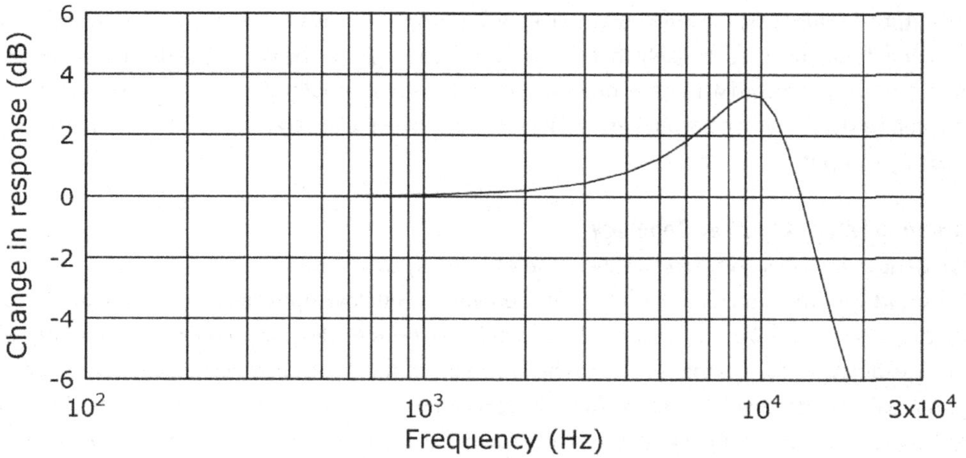

Figure 3 Variation in cartridge/preamp frequency response for Audio-Technica 150MLX cartridge between optimum load and 500pF input capacitance.

So... we not only have to get RIAA correct within the preamp, we need to keep input capacitance low to maintain a flat frequency response. If a particular cartridge needs to see a higher input capacitance, that's easy to add. Low input capacitance gives us the greatest flexibility for cartridge choice.

Issue 3: Overload Margins and Recovery

The lack of well-defined maximum signals coupled with the velocity characteristic of magnetic cartridges (whether moving magnet or moving coil), i.e., the output proportional to stylus velocity which in turn is proportional to frequency, is complicated by the nature of mechanical noise in the playback process. Ticks, pops, and mistracking are generally high frequency phenomena (or at least have a lot of high frequency energy in their spectral content), and thus can yield cartridge output signals orders of magnitude higher than the "nominal" maximum output, ill-defined as that is. Over the years, various measurements indicate that noise signals can easily exceed 20dB over nominal, with some claiming even higher numbers. Program signals at 10dB over nominal are not unknown.

On the bright side, the worst problems tend by definition to be transient events- if the top of a pop or tick is cleanly shaved off, that is of no great sonic consequence. However, if the recovery of the preamp following the overload isn't extremely fast, a transient event will manifest as some seconds of "choking," which is certainly of sonic consequence. The "choking" is generally a result of so-called blocking distortion, a well-known phenomenon in tube amps and other RC-coupled circuits. So the preamp either has to have overload margins that are well beyond any possible transient level or have instantaneous recovery afterward, or (preferably) both. 22dB above 5mV (the typical MM nominal output) is 62mV. For a safety margin, I am setting the target a bit higher at 100mV. The input stage must be able to handle that signal, and then be able to swing at least 100mV times the stage gain at its output. Fortunately, the RIAA curve will knock down the voltage presented to the second stage at higher frequencies, so it's likely that the HF overload margin will be limited by the first stage. One might argue that record warps and other subsonics have high amplitudes, a serious issue where the preamp has the highest gain, but here we are rescued by their relatively low velocities (and hence low cartridge output).

Bringing Things Together- Topology

We've now got a design target specified, and we're ready to put the pieces together.
Let's start with the requirement for balanced operation with low input capacitance. The first circuit that comes to mind is a simple differential amplifier (**Figure 4**). Referring back to our input capacitance estimates, this cuts the problem in half since the input capacitances of each half of the differential pair are effectively in series. This solution carries a 3dB noise penalty, but if the design is adequate, that 3dB is buried under the cartridge noise. So, a differential pair of 12AX7 would likely have an input capacitance of slightly under 100pF, a great improvement, but not quite low enough. Additionally, the higher intrinsic noise of the 12AX7 type makes that 3dB noise penalty more likely to be audible. A differential pair of ECC88 do quite a bit better- one could get the capacitance down to 50-60pF, which is a great improvement, but still not quite good enough.

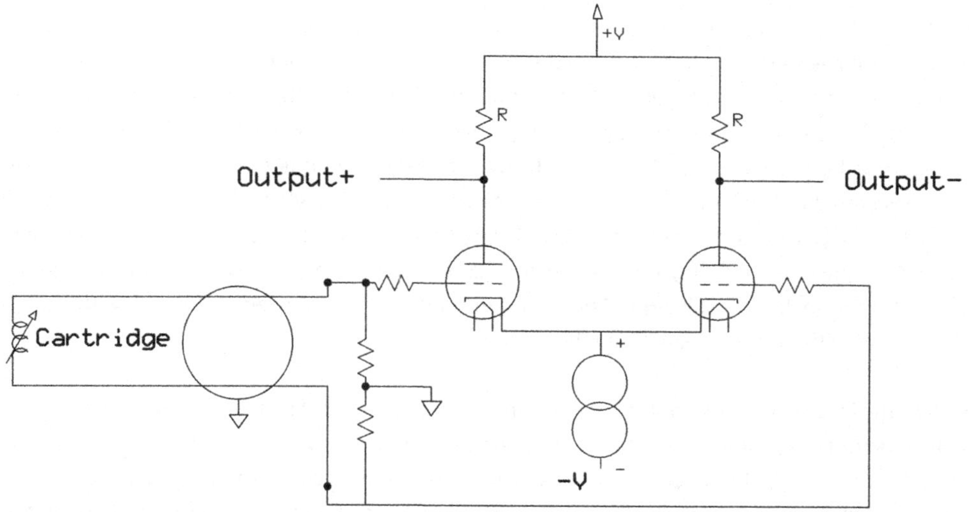

Figure 4 Standard tube differential amplifier. The use of a CCS in the cathode optimizes CMR and balance.

One trick from the RF world can be applied- **Figure 5** shows a pair of "neutralizing" capacitors crossed between sides. If they are sized and tuned properly, they will completely cancel the Miller effect, which dominates the input capacitance. An early prototype used this trick, but I found that the size of the capacitors was critical and varied from layout to layout and tube to tube. This necessitates trimmer caps, and the adjustment is tedious, tricky, and not terribly stable. Nice idea, but not practical.

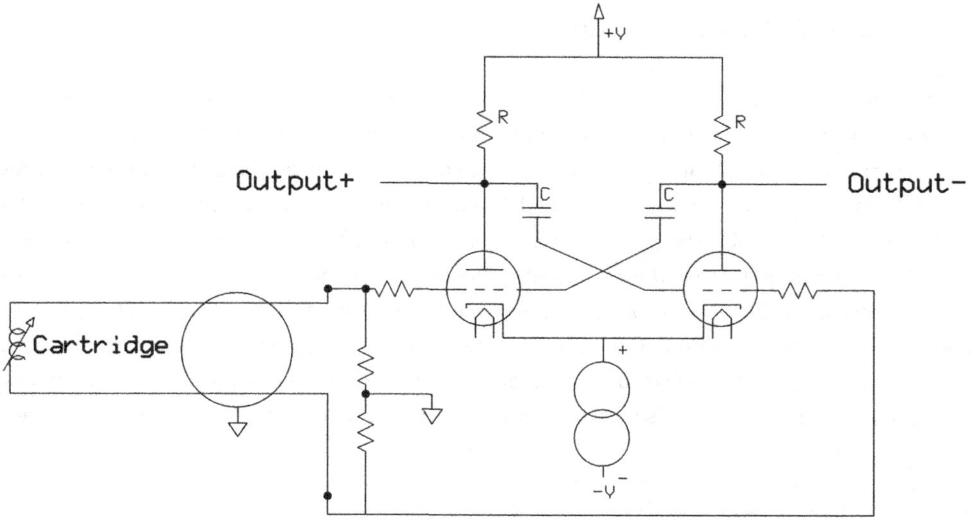

Figure 5 The use of neutralizing capacitors to decrease the input capacitance of a differential amplifier. The capacitors are generally equal to the grid-plate capacitance of the tubes.

Another way to drop the Miller effect is the use of cascodes. In a cascode circuit, the plate voltage of the input tube is held fairly close to constant, so the Miller effect is minimized. This can be an excellent solution and I would not argue with the choice to follow that path, but the practical implementation of cascodes in low noise circuits is fraught with problems. Power supplies for both the plate voltage and the grid voltage of the upper tube must be exquisitely quiet, since the power supply rejection is pathetic. A quick read of "The Tube Preamp Cookbook"[7], which is in reality the Complete Guide to Tube Cascodes, demonstrates the difficulty of getting these circuits stable and quiet. Eventually, Wright resorted to using a FET for the lower device for low noise, which somewhat violates my esthetic criterion, but it's still a good engineering solution that I've used in the past with success (though not without some tears along the way).

Back in the late 1970s, a brilliant young designer by the name of David Berning developed a very neat workaround to eliminate the Miller effect in triodes. His phono input circuit is shown in **Figure 6**[8]. In this circuit, the FET is a p-channel device used as a source follower to drive the cathode of a grounded grid amplifier. The idea was an excellent one, but certain aspects of the implementation caused the noise performance to be rather dreadful. First, the cathode/source bias resistor forms a voltage divider with the looking-in impedance at the cathode. This impedance is roughly the plate load plus plate resistance divided by the tube mu (actually mu + 1). For this implementation, the plate resistance is about 70k, the plate resistor is 110k, and mu = 100, so the cathode impedance is 1.8k. The gain at the cathode is about 60. The FET used has a transconductance of ~2mmho, so the source impedance driving the cathode is 500R + 2k7 =3k2. We've lost 2/3 of the input signal before it ever reaches an amplification device. With a FET e_n of 60nV/rt Hz, the signal isn't all that quiet to begin with. The remarkable thing about this stage is that it performed as well as it did[9]. Note that the stage, as drawn, is a non-inverting amp- one can use the grid of the tube as an inverting input, but it will of course have Miller input capacitance.

This topology can be greatly improved. First, we can use quieter FETs and tubes. The Toshiba 2SJ74 has always been the gold standard for low noise p-FETs, although it has sadly been discontinued; buying them on the secondary market is risky, since there has been a proliferation of fakes. Fortunately, Linear Integrated Systems has stepped into the breach and announced the availability of a 2SJ74 clone called LSJ74[10]. The LSJ74 boasts a typical input voltage noise density of 0.9 nV/rt Hz, more than 60 times better than the 2N5460 used by Berning. We can also clean things up by using a quieter tube than a 12AX7- there are quite a few good choices here, and for this circuit implementation, I have chosen the 6DJ8/ECC88/E88CC or one of its many other variants. The choice is based on easy availability, relatively low cost, and a long track record in test equipment and high quality audio applications. One could also consider other low noise triodes like D3a with even better per-

[7] Allen Wright, "The Tube Preamp Cookbook" 2nd Edition, 1997.
[8] David Berning, "Audio Amplifier," US Patent 4,163,198, 1979.
[9] In Berning's defense, low noise p-FETs were not common items in the 1970s.
[10] Information at http://linearintegratedsystems.wordpress.com/2013/09/23/an-important-announcement-concerning-the-lsj74-p-channel-single/

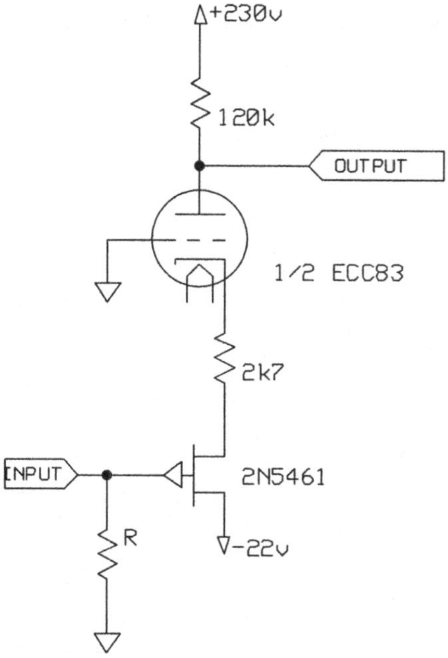

Figure 6 The input stage for the Berning TF-10 preamplifier. Note the use of a grounded grid amplifier to remove the Miller capacitance.

formance, but there's a correspondingly greater expense associated with that.

The input signal loss can be minimized by either bypassing the cathode-source bias resistor or my preferred approach, the use of an LED. A forward biased red LED will typically have an impedance of a few ohms, noise in the 0.4nV/rt Hz range, and nearly instantaneous recovery if momentarily underbiased by a clipped stage. What's not to like? The only caveat is that the voltage drop across an LED can only be chosen at specific discrete values, viz., 1.2V for an IR LED, 1.6V for a red LED, and so forth. We can gild the lily and reduce the signal loss to near zero by loading the plates of the ECC88 with a constant current source (CCS), which has extremely high impedance, simultaneously reducing distortion from the FET, the tube, and the LED. As a bonus, the power supply noise rejection is improved by a factor of approximately Z_O/r_p, where Z_O is the CCS output impedance (easily 1M or higher) and r_p is the tube's plate resistance, 3k or so for an ECC88 at 8-10mA standing current. The improved Berning topology is shown in **Figure 7**.

Even though there's both inverting and noninverting inputs, the balance and CMR are quite poor because of the differences in input impedance at each of the input ports. These can be optimized by reconfiguring things a bit. **Figure 8** shows the new topology, a fully differential circuit, which (about ten years ago, in a fit of whimsy about fancy names given to simple circuits) I named the SYclotron. The original application was as an input stage for power amps,

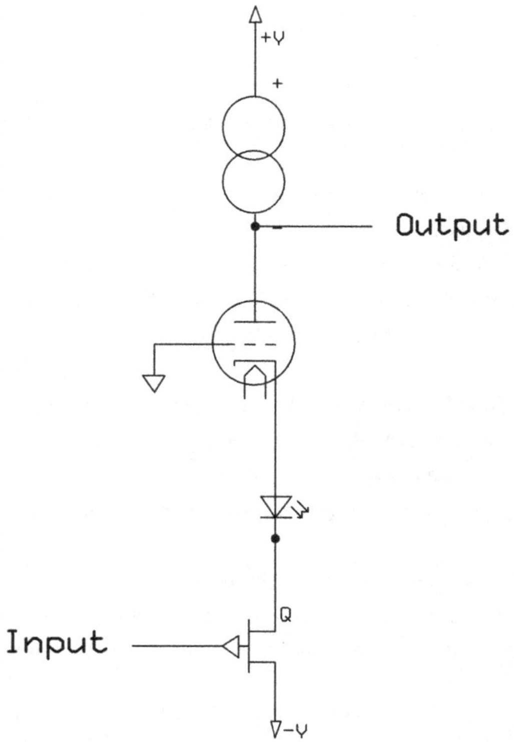

Figure 7. A modernized version of the Berning circuit, with improved gain, noise, and linearity.

where I have been using this basic topology for the past 30 years or so. When the need for an MM preamp came along, this seemed like an obvious choice.

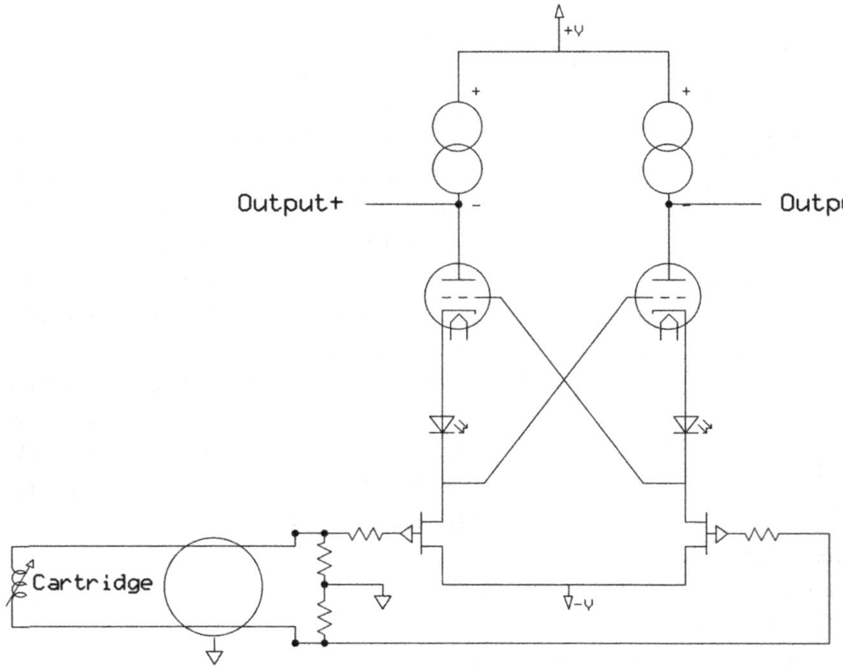

Figure 8 The so-called SYclotron circuit, the basis for the Equal Opportunity preamp.

The SYclotron is, at its heart, the same topology as the old "biological" amplifiers of the 1950s[11]. In those circuits, later popularized in the Audio Research line of amplifiers, a cathode follower simultaneously drove the grid of one side of the circuit and the cathode of the other. A second cathode follower performed the inverse of this task. The Miller effect is still present, but minimized due to the low source impedance of the cathode followers.

The p-FETs offer even lower source impedance and significantly lower noise as followers. Their opposite polarity to vacuum tubes means that the biasing is greatly simplified as well. Referring back to Figure 8, we can also see that we have incorporated the LED biasing trick; the original SYclotrons used the same resistor scheme as the Berning circuit.

The original biological circuit, with its cross-coupling, was designed to maximize CMR, and the SYclotron retains that virtue. It's clear that common mode signals are presented equally to the grids and cathodes of the amplifying tubes, so the common-mode gain is extremely low, about 1/mu. Con-

[11] See, for example, Stanley White, "The White Powertron Amplifier," Audio, Nov. 1953, reprinted in "Audio Anthology" vol 3, Audio Amateur Press 1990 for an early use of this circuit in audio.

trariwise, the differential gain is quite high, approaching 2 times mu. This implies better than 66dB CMR with this particular tube choice. A higher mu tube will do even better.

Now, let's step through and see where this circuit stacks up for noise. The input FETs each contribute 0.9nV/rt Hz of noise density, in series with the LEDs, which each contribute 0.4nV/rt Hz. The noises are uncorrelated, so add as power. So the total noise voltage density before the signal hits the tubes is $\sqrt{(2*(0.4)^2 + 2*(0.9)^2)} = 1.4$nV/rt Hz, referred to the input. The cartridge's noise density ranges from 5-25 nV/rt Hz according to the cartridge noise calculator referenced earlier, with the lower values at lower frequencies and higher values at higher frequencies, as expected. Since again the noises add by power, the increase in noise over the "bare" cartridge is negligible even at the lowest frequencies.

Estimating the tube noise is tricky since the variation from sample to sample is so high, and the 1/f corner of the noise characteristic is all over the map[12]. Thus, the given equivalent noise resistance specification on the datasheet is not terribly useful. If we take 1k of equivalent noise resistance as a guess, this translates to 4nV/rt Hz. At low frequencies, the noise increases over a "bare" cartridge from 5nV/rt Hz to 6.4nV/rt Hz. At higher frequencies, where our hearing is a bit more sensitive, the difference is lost in the noise, as it were.

One more significant noise source is the anode load. In this case, we used a CCS, which has an associated current noise density, i_n. This current noise interacts with r_p to produce a noise voltage at the plate $i_n r_p$. Referred to the input, this noise becomes $i_n r_p$/mu. The reported i_n of the CCS used in this circuit[13] is 14pA/rt Hz, r_p of an ECC88 is roughly 3k, and mu is 33 (thus differential gain is 66), so the input referred noise density from this source is 0.65 nV/rt Hz, again not a significant contributor. Other CCS circuits may give more or less noise contribution. A really outré way of accomplishing the same thing is to use a centertapped inductor as the plate loads, but that's out of my price range.

We seem to have gotten past the first stage unscathed by noise. Returning to overload, the stage gain of 66 means that we have to swing 6.6V differentially at the target 100mV maximum input, or 3.3V per side. Suffice it to say that overload margin will not be a problem!

The next question is the RIAA network. In previous work, Linear Audio's kind Editor and I showed that for a reasonable choice of resistance and capacitance values, the noise is totally negligible[14] compared to the signal amplified by the first stage. For this preamp, we will build a balanced differential version of the same RC topology, which only costs us one extra resistor.

Frequency response is a different matter. We want it to be stable against aging of the tubes, tube changes, selection of FETs, and anything else which might affect the source impedance of the first stage. For the tube and FET choices here, the source impedance is likely to be in the area of 2-3k from

[12] Burkhard Vogel, "The Glowing NoiseMaker- On the Demystification of Triode Noise," Linear Audio Vol 4, September 2012.
[13] Walt Jung, "High Performance Current Regulators Revisited," AudioXpress April 2009, 40.
[14] Stuart Yaniger, ibid.

each side of the amplifier, but that can vary quite a bit with tube changes and aging. **Figure 9** shows the change in frequency response from a source impedance increase of 2k. It's not dramatic, but it's likely audible as a slight dulling over time. To eliminate this potential effect, we can buffer the output of the first stage, but it's your choice- if you can live with the idea that your EQ might drift slightly but would rather eliminate a stage, you can skip the buffer.

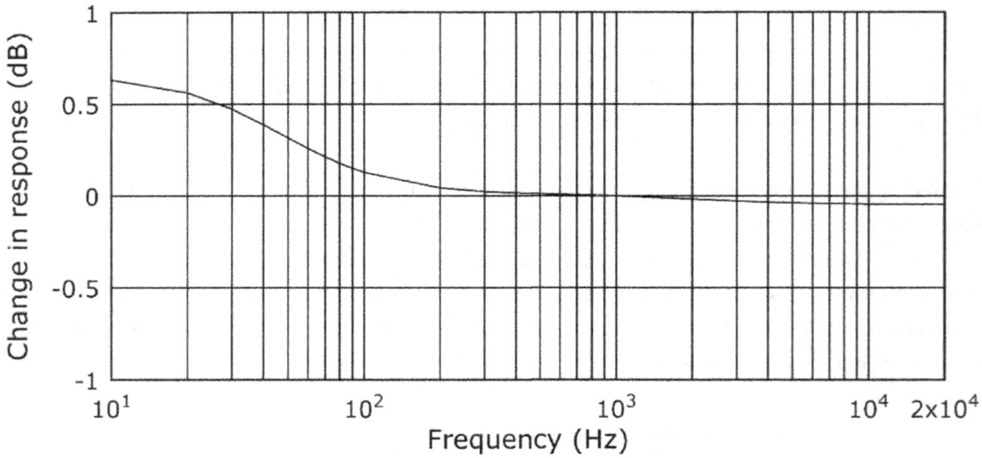

Figure 9 Variation of the preamp frequency response with the suggested RIAA circuit as the input tube plate resistance increases with age.

Of the various ways to implement the buffers, the two that immediately suggest themselves are cathode followers using triodes and source followers using MOSFETs. The former have a certain esthetic appeal and very high performance (assuming high transconductance triodes are used), but direct coupling to them from the plates of the first stage means that we must pay special attention to the heater supplies to insure that the maximum heater-to-cathode voltage rating isn't exceeded.

Surprisingly, MOSFET source followers have even better performance than cathode followers with respect to distortion and source impedance, even at the low-ish currents appropriate for this circuit[15]. Besides having a high voltage rating so they can be run off the same power rails as the tubes, the important specification for these MOSFETs is a low C_{rss}, since this represents the vast majority of the capacitance that would be seen by the tube stage (the gate-source voltage is close to constant, so that capacitance need not be charged and discharged). 500V MOSFETs having C_{rss} of 10pF or less are readily available and cheap; the pole created by the tube's source resistance and 10pF approaches 10MHz, so is of no real concern. The 100% feedback inherent to source followers and the relatively low voltage swings required mean that the linearity will be superb. As we will see, measurement bears this out. One unspecified parameter for MOSFETs is noise, so you might want to try a few different types. The specified MOSFETs worked fine for the prototype, and maybe they're all like that, but you can't count on it with a different lot. "Trust but verify."

[15] Morgan Jones, ibid.

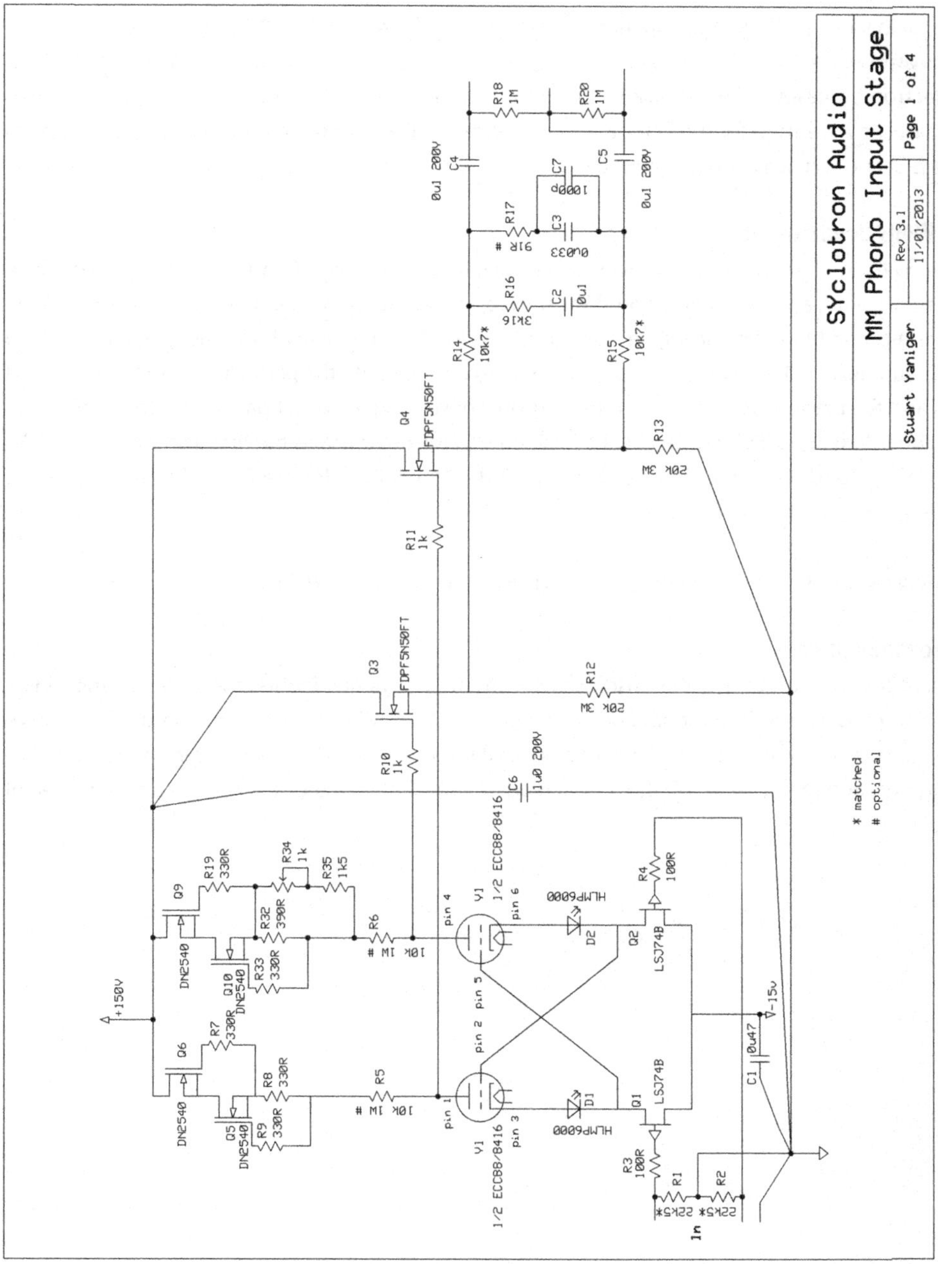

Figure 10 Signal circuit for the input stage of the Equal Opportunity. The second stage is identical, but with the passive RIAA network removed from the output.

The Second Gain Stage

Given the fine potential performance of the first stage gain block, there's no particular reason to avoid using it again for the gain block following the RIAA stage, although one could just as easily use a plain vanilla differential amplifier. For this section, the MOSFET buffers are not optional, since the preamp will be loaded with a cable and the input impedance of a line stage or (in my specific case) an electronic crossover. This choice is purely for convenience- we can use the same circuit boards and parts as were used for the first gain stage. Certainly, the overload performance and maximum voltage swing will be more than adequate.

H.G. Wells Department

In things to come (Part 2 of this article), I will delve into the details of part choices, matching, construction, layout, and grounding, and show the recommended power supply circuits. As mentioned before, a circuit board layout will be made available for the basic gain block plus RIAA filter and the power supplies. I will also present measurements to demonstrate the performance of the individual parts and the preamp as a whole. For the purposes of starting to collect parts or for those who love breadboarding and playing with circuits (who, me?), **Figure 10** shows the first gain block plus RIAA filter. The second gain block is, as just mentioned, identical but for the RIAA filter (which will be jumpered out on the circuit boards).

You now have a six month head-start before the next part of this article.

Acknowledgements

As usual, my ideas took form after discussions with smart guys, and in this case, besides Jan Didden, the smart guys were Morgan Jones and Kevin Kennedy. Murray Zeligman taught me about the Berning circuits and continues to be an inspiration. Last but not least, the Linear Systems folks were nice enough to send me a pile of p-FETs to try. If readers show interest, they may even do another run of the LSJ74s.

A Poor Man's Measurement Amplifier (pmma)

Burkhard Vogel

An experimental study on a low-cost balanced measuring solution that is capable of measuring - with smallest error rate - extremely low signal-to-noise ratios and output noise levels of amplifiers.

1. Introduction

In the last decade, a growing number of consumer audio products with balanced input/output capabilities came to market. For me it is unquestioned that balanced line connections are superior to un-balanced ones. That is why I am steadily increasing my measurement and audio equipment towards a balanced environment. One of the results will be a fully balanced-in/balanced-out MC phono-amp with passive equalization, based on solid-state and valve material, and without any detour to the un-balanced world. I hope to publish this design in a future issue of Linear Audio.
Here, in this study, I will concentrate on the measurement issues.

The measurement of noise in the balanced world requires special measurement amplifiers (MA). The best case would be one with balanced input and balanced and/or un-balanced output, being capable to feed one of the many computer based measurement solutions (CMS) that are affordable for the DIYer. Then, in this kind of environment, the task of the MA is to lift the Device Under Test (DUT) noise level far above the input noise level of the CMS. Finally, the CMS reading (in -dB) minus the MA gain (in dB) will show the output noise level of the DUT. This method allows the measurement of signal and noise levels in the region of 1µV_{rms} balanced, a low-level region nowadays mainly occupied by very high priced audio analyzers from eg. Audio Precision or Rohde & Schwarz.

Recently, in Stereoplay 2013-03, I read an article on the "NAD M51" DA converter. For me the most astonishing thing was the measurement result of the A-weighted output referred signal-to-noise ratio (SN): -122dBV(A) - and hum-free. Based on the FFT diagram in the review the noise curve shows white noise behaviour with a rather low 1/f noise corner frequency of approximately 350Hz. I have transferred the Audio Precision SYS 2722 measurement graph into the rough representation of **Fig. 1**.

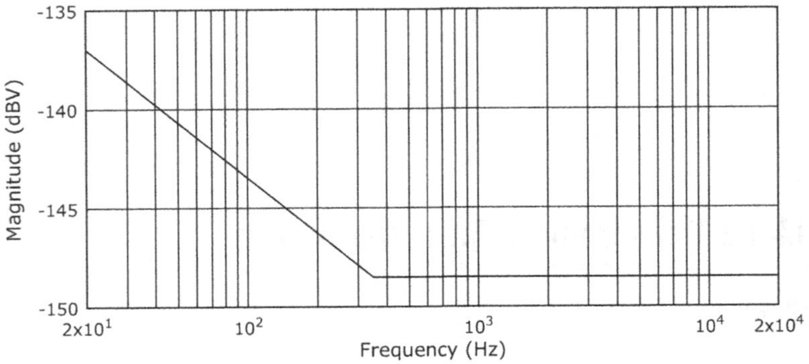

Figure 1 Output voltage noise density curve of the NAD M51

The 350Hz corner frequency 1/f-noise effect worsens the otherwise nearly pure white noise signal in the audio-band 20Hz to 20kHz (B_{20k}) by appr. 0.5dB. Additionally, the A-weighting effect counts for another 2.05dB. Theoretically, the corresponding purely white noise output signal would thus become -122dB(A) + 2.05dB - 0.5dB = -120.45dBV (NAD data sheet spec: -125dB ref. $2V_{rms}$≡ -119dBV, incl. 1/f-noise, of course). Measured in B_{20k} this extremely low SN level stands for an average output noise voltage density of 6.72nV/rtHz. The balanced output resistance shows a measured 141Ω (110Ω).

Nevertheless, this measurement result is not the truth! The above-mentioned SN of -122dBV(A) includes a certain portion of AP's unweighted input referred $SN_{i.AP}$ = -122.4dBV of the channel that was used for the measurement. With balanced inputs shorted, Stereoplay's four channels in two AP SYS 2722 analyzers, all equipped with 20kHz brick wall S-AES17 low pass filters, show input referred SNs between -122.0dBV and -122.4dBV (≡ -119.0dBu ... -119.4dBu in B_{22k} = 22Hz ... 22kHz with 3rd order low pass à la IEC 468).

The identification of a $1\mu V_{rms}$ signal spike in an FFT diagram seems not to be a problem, as long as the analyzer's FFT resolution can be set to eg. ≤ 5Hz. Depending on the sample rate and FFT size setting and as long as the analyzer's input referred noise voltage density is low enough (practically without impact on the measured signal) the representation of a B_{20k} $1\mu V_{rms}$ white noise signal on the screen will show a noise level far lower than -120dBV. In any case, it should be lower than an average level of 7.07nV/rtHz. Why? See further down in Section 2.

I do not want to debate what is the 'right' lowest noise level and thus the highest possible SN value of HiFi equipment. In Douglas Self's L|A Vol. 5 article, Table 4, we can easily study what we can achieve in that field. Because of their extremely low value all measured figures had to be corrected according to the AP noise impact. To get dBV in B_{20k} simply subtract 2.22dB + 0.81dB = 3.03dB from the table's dBu values; here, 0.81dB comes from the reduction of the bandwidth (0.41dB) plus the low pass change (0.4dB) from 3rd order Butterworth to the brick wall version.

Thus, I guess an amplifier output referred SN benchmark of -120dBV in B_{20k} is still an extreme value and as music listener, I can live with it - even in the quietest moments. However, seen from a development point of view, we have some open questions on an MA that should be capable to measure such extremely low SNs with lowest possible impact on the measurement result. With that in mind, with this study I will try to answer at least the following questions:

1 What should be the level of input noise voltage density of an MA with balanced input that causes the low output noise voltage density of a DUT with a measured $SN_{o.DUT.m}$ = -120dBV/B_{20k} to be off by eg. 0.1dB maximum only?

2 How does the DUT's output resistance additionally worsen the measurement result and what are the roles of the MA's input resistance R_{in} and its input noise current density $i_{n.i}$?

Based on the the follwing three subgoals how could a rather low budget and easy to built measurement-amp look like?

3.1 Flat frequency and phase response in B_{20k} (± 0.02dB / ± 10°)
3.2 Overall gain: 1000
3.3 Balanced input and balanced/un-balanced outputs

How can we calculate the noise performance of that new MA?

We will see further down that the chosen pmma concept fulfils all above-mentioned goals. In addition, it costs less than 150 EUR (2013).

2. The Input Noise Voltage Density Question

Before we start developing the MA and as consequence of the first question and its -120dBV goal, in order to find the MA's input referred noise voltage density allowed ($e_{n.i}$) we have to go through some math first. The equations for the required MA input referred $SN_{i.ma}$ look as follows:

With the DUT output $SN_{o.DUT.m}$ [dBV] < $SN_{i.ma}$ [dBV] and with

$$B = SN_{o.dut.m} - SN_{i.ma} \quad [dB] \tag{1}$$

we obtain the MA's allowed input referred $SN_{i.ma}$ as function of B as follows:

$$SN_{i.ma}(B) = 20\log\left(10^{\frac{SN_{o.dut.m} - B}{20}}\right) \quad [dBV] \tag{2}$$

Substitution of (2) in the following equation will lead to the expression for the allowed impact $W_e(B)$ [dB] of the MA on the DUT. It should become - in our case here - 0.1dB.

$$W_e(B) = 20\log\left(\sqrt{10^{\frac{SN_{o.dut.m}}{10}} + 10^{\frac{SN_{i.ma}(B)}{10}}}\right) - SN_{o.dut.m} \ [dB] \qquad (3)$$

Fig. 2 shows the corresponding graph (see its derivation on the separate 1st Mathcad worksheet, WS-1, on the L|A website).

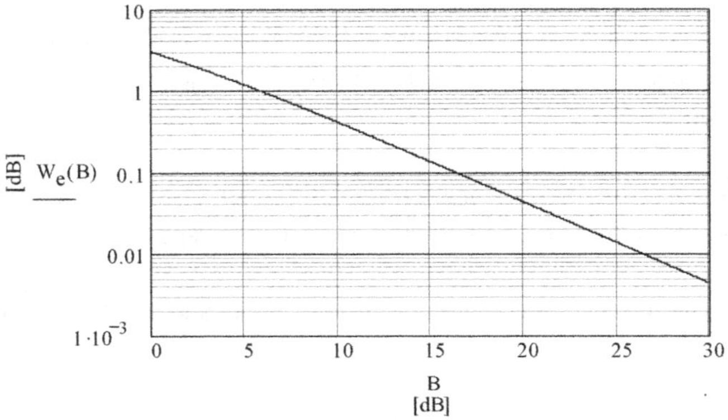

Figure 2a Worsening Figure $W_e(B)$ as function of the difference B of two SNs.

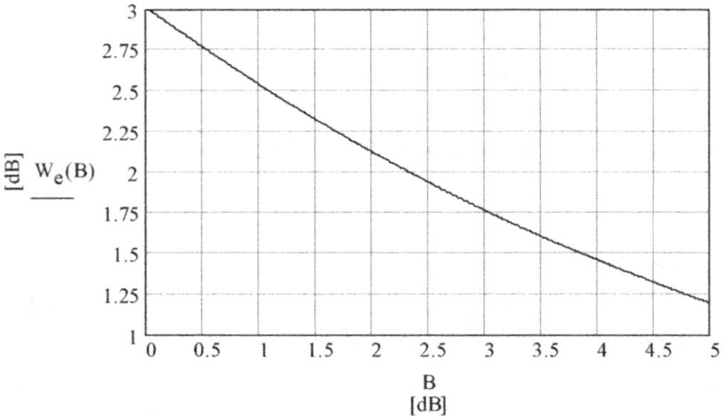

Figure 2b Zoomed version of Fig. 2a for correction purposes of two SNs with values that are close together (B ≤ 5dB).

Based on the calculated range of Worsening Figures (y-ordinate), at the allowed Fig. 2a value of $W_e(B)$ = 0.1dB we can now pick-out the delta B (x-ordinate) between the DUT's output referred $SN_{o.DUT}$ and the $SN_{i.ma}$ of the MA: it becomes 16.33dB. Hence and in other words: to ensure a ≤ 0.1dB MA impact down to a DUT output SN level of -120dBV on all kinds of SN measurement in B_{20k} we need an optimal MA input referred $SN_{i.ma}$ of -136.33dBV = -120dBV - 16.33dB. Thus, transferred into a white noise voltage density level flat over the complete audio band we need an MA that offers at least a balanced average input noise voltage density ≤ 1.08nV/rtHz, hence, without 1/f-noise corner frequency in B_{20k}.

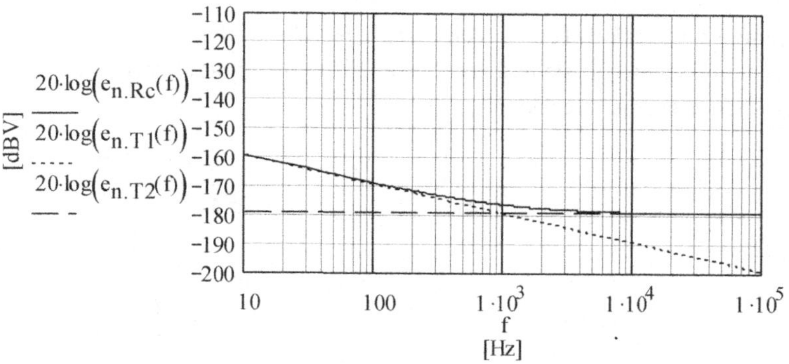

Figure 3 Noise voltage density situation with a corner frequency of 1kHz (incl. tangents).

If we would, for example, allow a 1/f-noise corner frequency of 1kHz, the average noise voltage density in B_{20k} would grow to -135.05dBV (see **Fig. 3** and 2nd Mathcad worksheet, WS-2) and the MA's impact on the DUT's $SN_{o.DUT}$ would also grow to $W_e(B) = 0.134$dB. If we would keep the original -136.33dBV level and we would accept the example corner frequency of 1kHz, then, to keep an average of 1.08nV/rtHz in B_{20k} the white noise region >1kHz of the MA's input noise voltage density has to drop from 1.08nV/rtHz to a balanced 0.93nV/rtHz.

Thus, to keep things on the safe side and as derived sub-goal for our purposes here the pmma should offer an input noise voltage density of ≤ 1nV/rtHz average in B_{20k} with no 1/f-noise corner frequency. The permitted amount of input noise current density is part of the next section's discussion.

3. The Roles of the MA Input Resistance and Input Noise Current Density

3.1 Influence of the DUT output resistance

I think it is obvious that a voltage divider effect between the DUT's total output resistance $R_{o.tot}$ and the MA's total input resistance $R_{in.tot}$ will lead to an improved meter reading of the output referred $SN_{o.DUT.m}$. This would happen if we would keep the reference level like the one without voltage divider effect.

However, if we would reduce the reference level according to the voltage divider ratio, the measured $SN_{o.m}$ of the DUT would become the DUT's real one. To get the real $SN_{o.r}$ with reference to the original reference level (in dBu or dBV) we have to compensate for this kind of erroneous improvement by simply adding the magnitude of the voltage divider gain-loss $G_{loss1.e} = 20\log(G_{loss1})$ to the measured $SN_{o.m}$. Hence, we obtain the real $SN_{o.r}$ as

$$SN_{o.r} = SN_{o.m} + |G_{loss1.e}| \tag{4}$$

In **Fig. 4** the DUT has two kinds of output resistances; R_{o1} and R_{o2} represent the real internal output resistance of the DUT. Not taking into account any external noise producing influences, we assume that the whole noise generation of the DUT already covers their internal electronic noise.

A typical example for this kind of output comes from the valve world: the anode output of a common cathode (or common grid) gain stage. In most cases, op-amp outputs offer an output resistance close to 0Ω, I guess in most cases < 10Ω in B_{20k}.

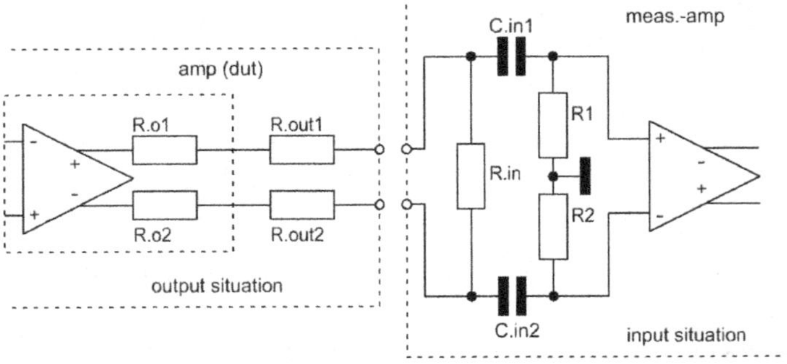

Figure 4 Left: general output situation of the DUT. Right: balanced input of measurement instrument.

By assuming the two input capacitances C_{in1} & C_{in2} shorted, the gain-loss $G_{loss1.e}$ becomes:

$$G_{loss1.e} = 20\log\left(\frac{R_{in.tot}}{R_{o.tot} + R_{in.tot}}\right) \tag{5}$$

$$\begin{aligned} R_{in.tot} &= R_{in} \| (R1+R2) \\ R_{o.tot} &= R_{o1} + R_{o2} + R_{out1} + R_{out2} \end{aligned} \tag{6}$$

Together with the following cable capacitance, the two resistors R_{out1} and R_{out2} play an additional low-pass role in cases of very low R_{o1} + R_{o2}. Many manufacturers put them in place in order to ensure a high-frequency cut far outside B_{20k}. In conjunction with the input noise current of the next stage, these DUT internal resistances may produce significant portions of additional noise.
Thus, both kinds of resistances play a role in the gain-loss but can also be significant producers of additional noise.

3.2 The input noise current density question

Fig. 5 shows the situation of the Fig. 4 based noise sources. Here, $e_{n.o.DUT.eff}$ is the DUT's noise voltage after the voltage divider, however without noise voltage impact of the MA input.
The noise voltage impact of the MA comes into the game in **Fig. 6**. It is composed of several sources, such as the Fig. 5 ones and those of a typical balanced input stage. Here, $e_{n.i1}$ and $e_{n.i2}$ are the input noise voltage densities of the MA's input stage, both of equal value $e_{n.i}$, series connected, thus increasing their value by √2. The input noise current densities $i_{n.i1}$ and $i_{n.i2}$ are both of equal value $i_{n.i}$. Also connected in series they decrease their combined value by √2.
Fig. 6 shows the result after connecting the DUT output with the MA input: all noise sources form what I call the total input referred noise voltage density $e_{n.i.tot}$ at the (+) and (-) input of the MA.

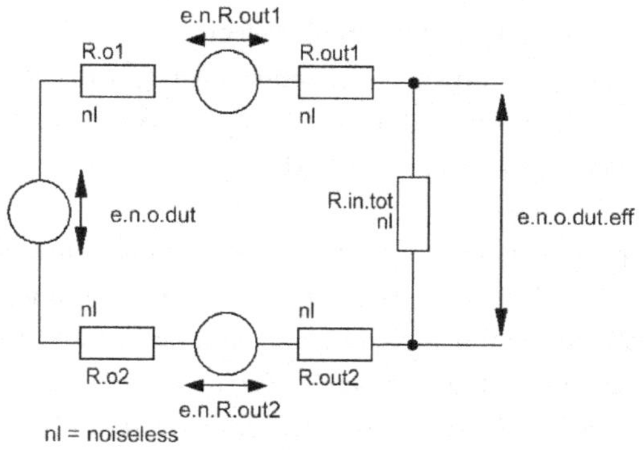

Figure 5 Effective output noise voltage of the DUT.

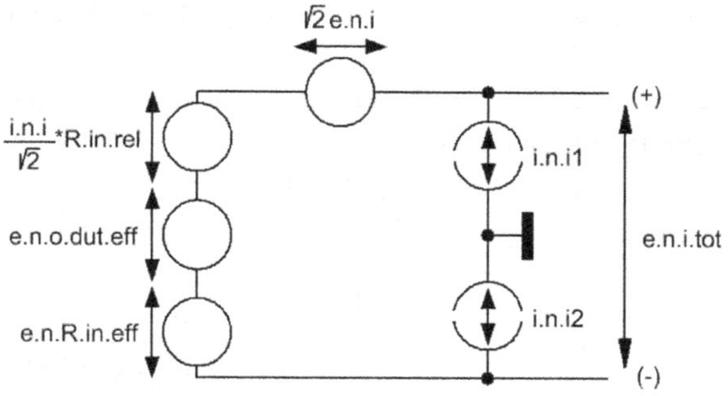

Figure 6 The creation of the total input noise voltage of the measurement amp.

For calculation purposes, we need the following equations:

$$e_{n.i.tot} = \sqrt{2e_{n.i}^2 + \left(\frac{i_{n.i}}{\sqrt{2}} R_{in.rel}\right)^2 + e_{n.R.in.eff}^2 + e_{n.o.dut.eff}^2} \tag{7}$$

$$\begin{aligned} e_{n.R.in.eff} &= e_{n.R.in.tot} G_{loss2} \\ R_{in.rel} &= R_{in.tot} \| R_{o.tot} \\ G_{loss2} &= \frac{R_{o.tot}}{R_{o.tot} + R_{in.tot}} \end{aligned} \tag{8}$$

In (7), 2nd term, we can see that, as a result of a noise voltage, the input noise current density flows also through the DUT output resistors R_{out1} & R_{out2} and resistances R_{o1} & R_{o2}. Here, their sum plays by far the major role because of the relatively high value of $R_{in.tot}$ in $R_{in.rel}$ (see (6) and (8)). According to Fig. 5 and with a noiseless $R_{in.tot}$ we can calculate the 4th term.

The inclusion of R_{o1} & R_{o2} into the 2nd term is questionable. Imagine an anode output of a triode with an anode load resistor R_a. Thus, in case of a bridged cathode capacitance the output resistance becomes $R_a \| r_a$ (r_a = internal triode anode resistance). I doubt that the input noise current of the MA will also flow through the triode against the triode's electron current from cathode to anode, hence, from anode to cathode. Consequently, in (5) we should take R_a as the value for R_{o1} and R_{o2} (in case of a balanced output stage). As long as the value of R_a is unknown, we can take the measured or data sheet indicated output resistances of the DUT.

I have calculated a graph that demonstrates the respective relationships. **Fig. 7a** shows the additional noise voltage creation if we measure a DUT with output resistances in the range up to 10kΩ. Starting point is 1nV/rtHz and I took three different input noise current density versions: BJT input devices with 1pA/rtHz (solid & dotted traces and subscript a), 2pA/rtHz (dashed and dashed-dotted traces and subscript b), and FET input devices with 50fA/rtHz (dotted and solid traces and subscript c). Additionally, I have chosen two different MA balanced input resistances: 18.182kΩ (pmma and subscript 1) and 200kΩ (AP and R&S and subscript 2). At the right-hand scale of Fig. 7a I have added the trace values for $R_o = 10k\Omega$. It should allow easier identification of the traces in the left graph (see also Mathcad worksheet WS-3).
Fig. 7b shows the zoomed version for the 18.182kΩ MA input resistance case, up to $R_o = 2k\Omega$ vs. the three different input noise current densities.

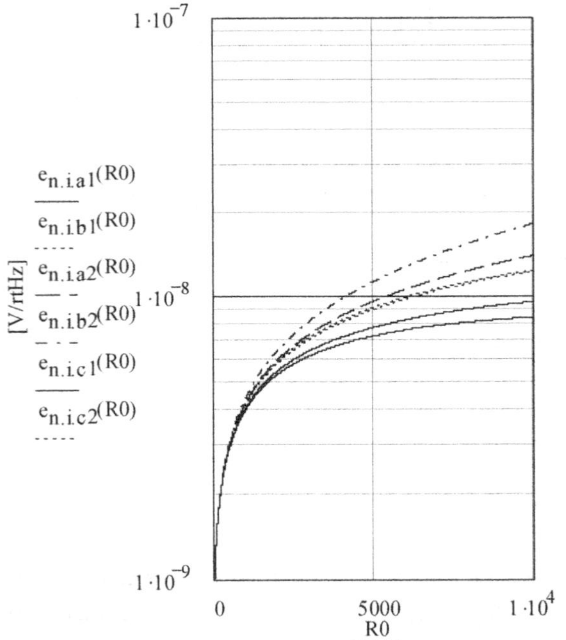

$$e_{n.i.b2}(10^4\Omega) = 18.24 \times 10^{-9}\,V$$

$$e_{n.i.a2}(10^4\Omega) = 14.022 \times 10^{-9}\,V$$

$$e_{n.i.b1}(10^4\Omega) = 12.378 \times 10^{-9}\,V$$

$$e_{n.i.c2}(10^4\Omega) = 12.304 \times 10^{-9}\,V$$

$$e_{n.i.a1}(10^4\Omega) = 9.527 \times 10^{-9}\,V$$

$$e_{n.i.c1}(10^4\Omega) = 8.367 \times 10^{-9}\,V$$

Figure 7a DUT output resistance vs. various MA total input noise voltage variants.

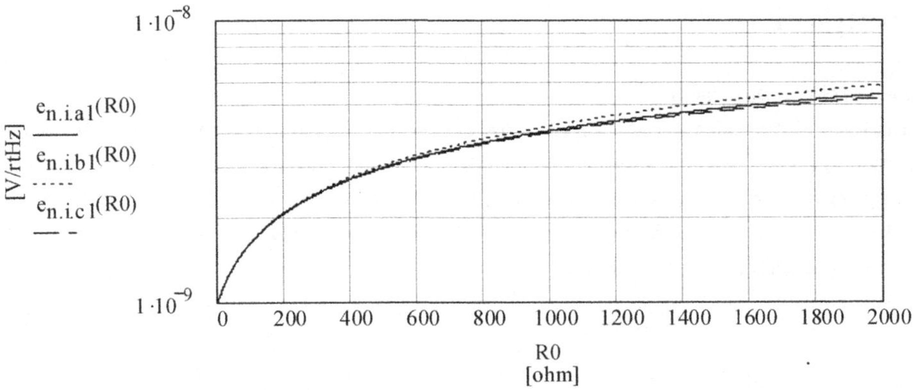

Figure 7b Zoomed version of Figure 7a (for $R_{in.tot}$ = 18.182kΩ only).

3.3 Conclusions

With an MA input resistance of 18.182kΩ a look at Fig. 7b signals that there is no big difference up to 2kΩ output resistance between nearly no input noise current and a rather high one of 2pA/rtHz. Additionally, for all shown Fig. 7a cases up to 2kΩ the input resistance difference does not play a major role either. However, we can see that 180Ω output resistance doubles the 1nV/rtHz (= -180dBV) input noise voltage density goal. With that we no longer can maintain the 0.1dB worsening goal for the measurement of a 1µV$_{rms}$ (= -163dBV average noise voltage density level) DUT output noise signal. With 2nV/rtHz (= -174dBV) the difference B to -163dBV is only 11dB; hence, as of Fig. 2, the worsening figure grows to 0.33dB. At eg. 1kΩ output resistance we obtain thus W_e = 1.2dB.

To keep both the complexity as well as the required budget of the whole MA circuit as low as possible (Occam's Razor!) I will adequately take the W_e effects into account and my input noise current density goal for the MA should be set to $i_{n.i} \leq$ 1pA/rtHz. In such a high gain and low-noise configuration FET input op-amps do not offer any advantage over their BJT input counterparts. Although their input noise current impact is marginal, the chance for a very much higher 1/f-noise corner frequency is a given fact. With that in mind the two chosen input noise density values are thus in line with the same values of the un-balanced MA that is already described in Fig. 22.1 of my book 'The Sound of Silence' [1], however, now in a fully balanced mode.

Nevertheless, as input devices for MAs with gains ≤100 and input noise voltage densities in the range of 3nV/rtHz … 6nV/rtHz, FET input op-amps can very well be used, like eg. those of the OPA627 or OPA827 family. Using those, noise voltage loaded outputs of valve gain stages with high output resistances > 10kΩ can easily be measured, with a very low MA noise impact on the measured signal. By changing the MA input resistance to 1MΩ, or even higher, such a gain stage could also be fit into the pmma configuration presented here, with only one balanced input stage.

4. The final pmma

4.1 Principal circuit approach

Low complexity also implies application of rather easy to handle circuits with op-amps or special measurement ICs. I do not know a single IC that offers the specs I want to have; hence, I took the traditional approach by arranging two input op-amps in a balanced in/out configuration and a third one that creates the un-balanced o/p. This is nothing else but the well-know instrumentation amplifier (INA), its principal circuit is shown in **Fig. 8**.

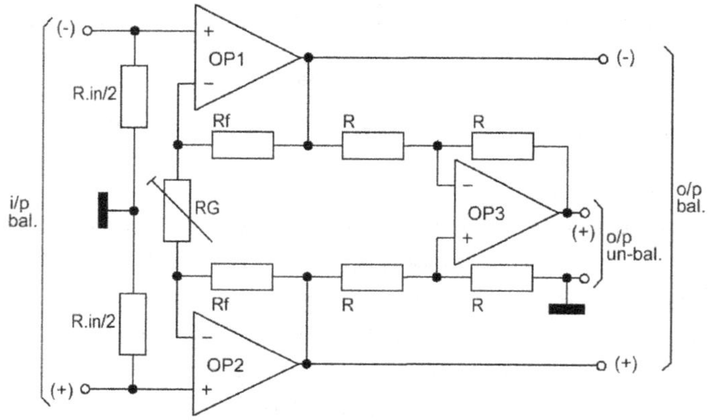

Figure 8 Principal INA circuit.

Unfortunately, to create 1nV/rtHz = $\sqrt{2}$*$e_{n.i}$ average input noise voltage density in B_{20k} I also do not know an op-amp that offers an average input noise voltage density of $e_{n.i}$ = 0.707nV/rtHz in B_{20k}. However, the only one I know that offers an input noise current density of 1pA/rtHz also offers 0.85nV/rtHz input noise voltage density in B_{20k}, the LT1028. To get a lower noise voltage density we must thus work with a paralleled input stage of four op-amps and two additional summing op-amps. Hence, the final MA consists of seven op-amps à la **Fig. 9.**

The resistors Rf and R should have 0.1% tolerance. With a variable RG, we can set each input stage to a gain of 500. OPs 5 & 6 sum-up their signals and we obtain thus a gain $G_{M.bal}$ = 1000 ≡ + 60dB at the balanced output. The common mode gain $G_{CM.bal}$ of that stage is 1, hence, the common mode rejection ratio CMRR of the balanced stage becomes (at least) 60dB = 20*log($G_{M.bal}$/$G_{CM.bal}$). OP7 serves as subtractor with a gain of 1. Consequently, the gain $G_{M.unbal}$ for the MA's balanced to un-balanced conversion becomes + 60dB too. Its CMRR offers an additional 60dB.

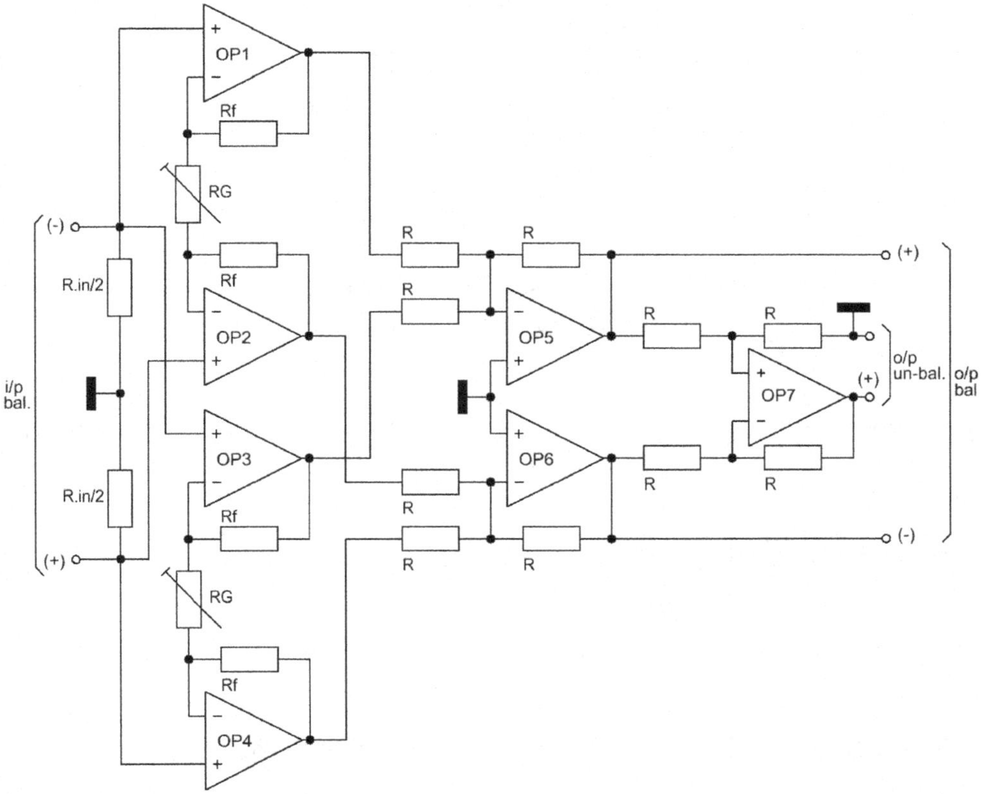

Figure 9 Principal final MA. (+) and (-) at the input and outputs indicate the phase relationships.

4.2 Noise calculations

Mathcad Worksheet WS-4 gives the full details of the calculation path. However, here are the basics:

4.2.1 Rule-of-thumb SN calculation approach

To get a feeling on the output referred SN we will start with a rule-of-thumb (ROT) approach. It is based on a single op-amp gain stage in series configuration with a gain of 1000 and input shorted. **Fig. 10** gives the details.

To get $SN_{o.ROT}$ in B_{20k} and with G_M =1000 the ROT math goes as follows:

$$SN_{o.rot} = 20\log\left(\frac{e_{n.o.rot}}{1V}\right) + 43\,dB \quad [dBV] \tag{9}$$

With the op-amp's $e_{n.i}$ = 0.85nV/rtHz and RG's $e_{n.RG.ROT}$ we obtain the output noise voltage density $e_{n.o.ROT}$:

$$e_{n.o.rot} = G_M\sqrt{e_{n.i}^2 + e_{n.RG.rot}^2} \tag{10}$$

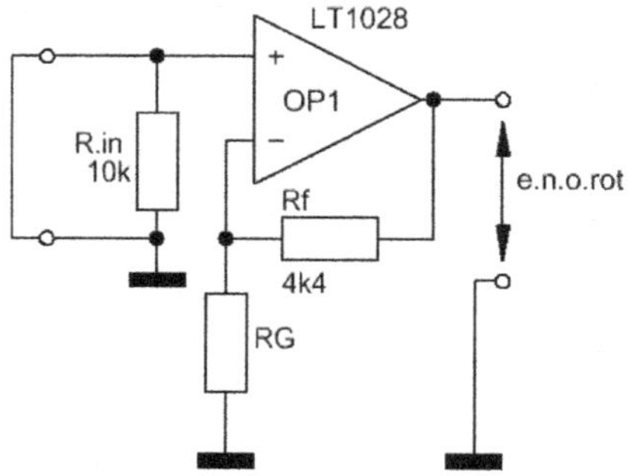

Figure 10 Circuit for rule-of-thumb SN calculation.

$$RG = \frac{Rf}{G_M - 1} \tag{11}$$

$$\begin{aligned} e_{n.RG.rot} &= 0.13\sqrt{\frac{RG}{\Omega}} \quad [nV/\sqrt{Hz}] \\ &= 0.273 \quad [nV/\sqrt{Hz}] \end{aligned} \tag{12}$$

$$e_{n.o.rot} = 893 nV/rtHz \tag{13}$$

Hence, $SN_{O.ROT}$ and the gain stage input referred noise voltage density $e_{n.i.ROT}$ become:

$$SN_{o.rot} = -78.0\,dBV \tag{14}$$

$$\begin{aligned} e_{n.i.rot} &= \frac{e_{n.o.rot}}{G_M} \\ &= 0.893\,nV/\sqrt{Hz} \end{aligned} \tag{15}$$

We have now found the two most important noise values that show where we should end-up with the pmma experiment.

4.2.2 Detailed SN calculation approach (details see MCD WS-4 on the L|A website)

The path through the noise calculation jungle of such a gain stage looks rather difficult. However, I will present it here because gain stages with lower gain than x1000 and other than the shown passive components bear many traps. I hope I have avoided most of them.

According to the Fig. 9 MA configuration, we have to split the calculation course into three different parts:

- Input stage with OP1 … 4 (= 1st stage),
- 2nd and balanced output stage with the two summing stages OP5 & OP6,
- 3rd stage with balanced to un-balanced converter OP7.

Normally, in a x1000 MA the influence of the noise of the 2nd and 3rd gain stage can fully be ignored; the 1st stage creates far more noise than the two other stages. However, in configurations with lower gain, the noise impact of the two output stages becomes significant. We should know the noise producing mechanism behind that fact. This fact is also the cause for the increase of the input referred noise voltage density of commercially available INAs, if we set them to low gains.

The noise model of one-half of the 1st stage with input shorted is given in **Fig. 11**. However, the mathematical treatment of the noise voltage of RG and its transfer to the output created some problems that I could not solve in an easy manner. I have been searching for an approach that would mirror reality equally well: **Fig. 12** shows the solution. Here, we split RG into two halves, $RG1_a$ and $RG1_b$. They produce easier to handle un-correlated noise voltages. We add an input load R0, and, as a reminder, we show the connections to the parallel operating stage with OP3 & 4.
The noise model of the other half of the input stage looks absolutely the same. By paralleling these two halves, the resultant output noise voltage becomes factor √2 higher.

Paralleling goes via the summing stages OP5 & 6. **Fig. 13** shows the corresponding noise model. In **Fig. 15**, we find jumpers1 ... 4. They can produce short-circuits at the inputs; hence, we can separately check the calculated output noise voltage of these two stages. Here, RP is the parallel configuration of the input resistors at the (-) input (R11||R12, R13||R14 in Fig. 15).
Finally, the well-known subtraction stage in an INA shows its noise model in **Fig. 14**. Here, RP is the parallel configuration of the resistors at the (+) input (R17||R19 in Fig. 15).

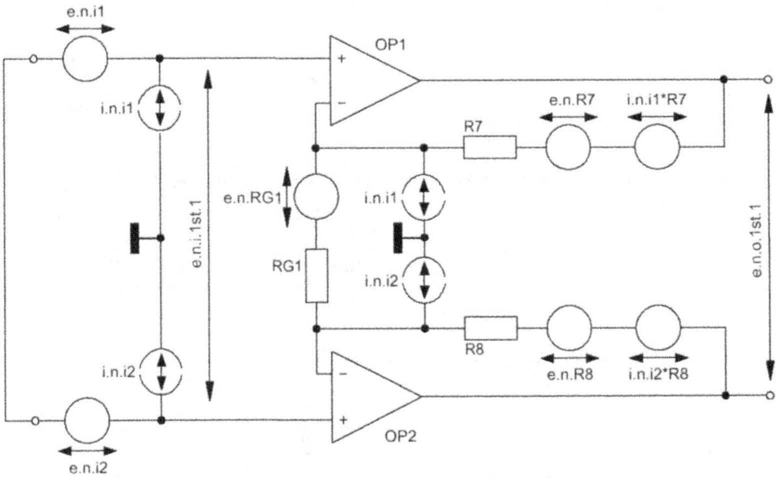

Figure 11 Noise model of one-half input stage (Op1 & OP2).

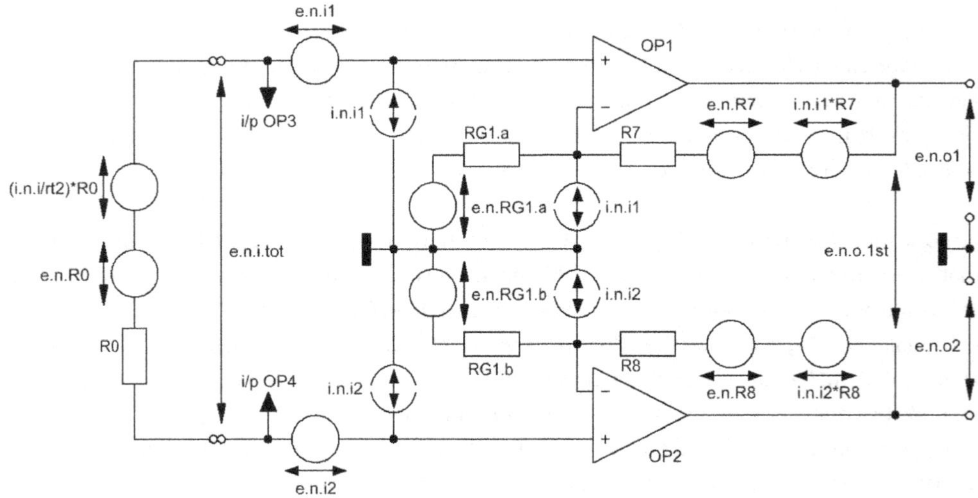

Figure 12 Re-designed Fig. 11 noise model for easier noise calculations

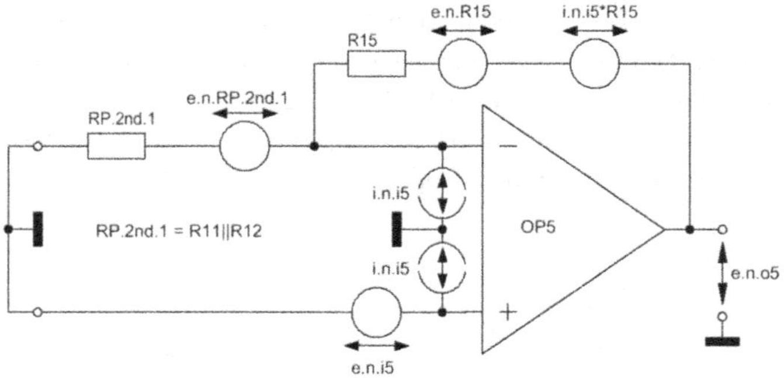

Figure 13 Noise model of one of the two summing stages (OP5 & OP6).

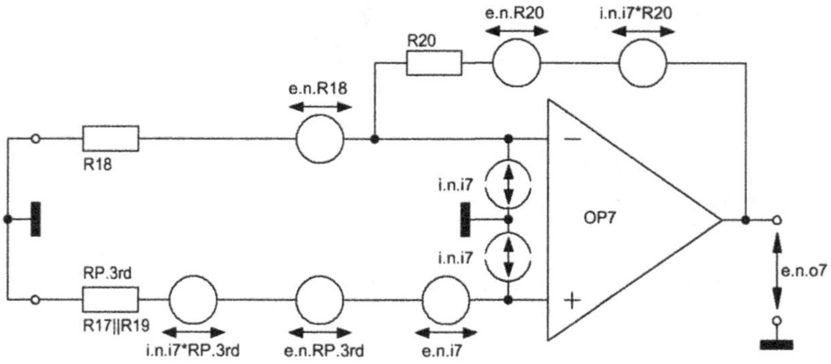

Figure 14 Noise model of the bal to un-bal converter (OP7).

I refrain from filling the pages with a huge amount of equations. The chosen process of output noise voltage calculations follows the block concept: each gain stage's output noise voltage density becomes multiplied by the gain of the following stages. Then, to get the total output referred noise voltage density $e_{n.o.tot}$, we sum-up the un-correlated output noise voltage densities. Division by the gain of the MA will lead to the input referred noise voltage density $e_{n.i.tot}$. In addition, the calculation course allows playing around with various values of input load resistances R0. To meet the goal from the beginning it should become ≤ 1nV/rtHz with R0 = 0Ω.
Worksheet WS-4's Fig. 7 graph of the input load dependent output referred SN shows rather drastically that it would be always better to measure DUT output SNs with DUT output resistances close to 0Ω, in any case ≤ 10Ω, to keep the pmma's average input noise voltage density ≤ 1nV/rtHz. For consumer products, this is rather seldom the case. The final SN determination should always consider these effects.

4.3 Complete pmma circuit

Fig. 15 shows the final and complete built-up pmma circuit. For a better understanding, the following list gives some hints:

1 The main balanced i/p is J01; it should be used for balanced purpose only. For other purposes, I added J02 & J03; they allow un-balanced application by eg. switching S2 and nulling J03. Usage of these inputs may lead to unstable conditions.
2 S1 represents the ground lift, thus, in most cases solving the *pin1 problem*; based on a Jensen Transformer recommendation, an open S1 and the sequence of R1 and C1 ensure grounding for very high frequencies.
3 R3 and R4 parallel to R2 set the input resistance R_{in} of 18.182kΩ. An attempt to increase R3 and R4 to 100kΩ each works too, however the offset-nulling process becomes a nightmare and after each input load change the amp's output shows a tendency to run into the power supply line level. This effect disappears after switching the psu off and on a few seconds later.
4 To allow direct amplification jumpers 5 & 6 can bridge the input Cs (C2a + C2b & C3a + C3b). I do not recommend it. It would require a new offset trimming process for each new input load.
5 For easy change to another configuration than the one shown, the two input stages OP1 & OP2 and OP3 & OP4 (including their feedback resistors) are built-up on two separate plug-in pcbs; RG examples for gains of x10 and x100 are given in **Fig. 16**
6 C4, C5, Cx, Cy may be a necessary addition to the circuit; it depends on the pcb layout as well as on the shielding effect of the case; however, in my configuration I only had to pack Cx and Cy on R7 … R10, each 22pF ceramic and selected 1%.
7 Setting jumpers1 … 4 to ground allows easy offset-nulling of OPs 5 … 7 before the offset-nulling process of the input stages should take place
8 A DC voltmeter between TP1& TP2 or TP3 &TP4 measures the alternately performed offset-nulling of the input stages by P3 … P6, hence, after replacing jumpers1 … 4 there should be no offset voltage left at outputs PL01 and J04. I must point out that offset-nulling makes sense only after a 15 min heat-up with shorted input, and without jumpers 5 & 6. It takes some additional

time to get to 0.00V at the outputs.

9 Fed by a 1kHz / 5mV$_{rms}$ balanced input signal and measured between testpoints T1 / T2 and T3 / T4 the gains of the two input stages must then be trimmed by P1 & P2 to 500 each and 2.5V$_{rms}$ output level; automatically, there will be an output signal of 5V$_{rms}$ at both outputs PL01 and J04.

10 In case of a x10 or x100 gain solution only one input stage will be necessary.

11 Replacement of R3 and R4 by eg. 1MΩ each and R2 by 2MΩ leads to an input resistance of 1MΩ, very useful for 1st gain stages with FET op-amps for the measurement of high resistive valve output stages.

12 I've set the input resistance to 18.182k, a value high enough to measure DUTs with output resistances up to 2kΩ balanced. Of course, the measured values have to be corrected for the gain-loss of the voltage to obtain the DUT's actual output referred SN.

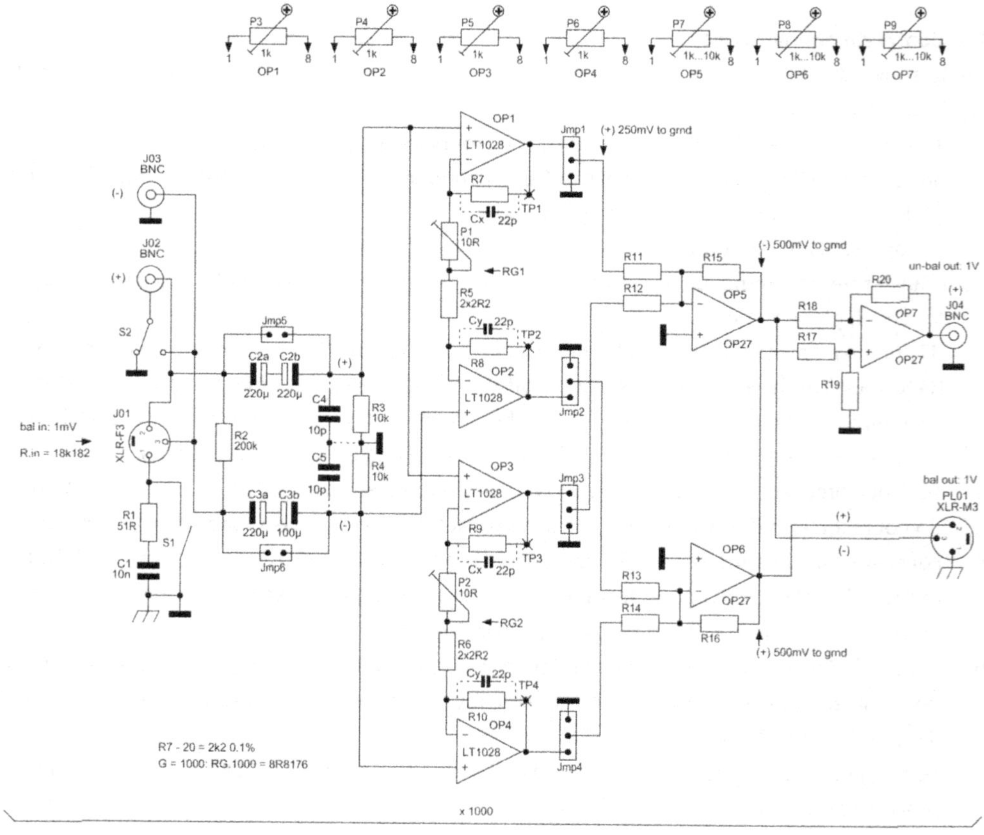

Figure 15 Complete MA circuit.

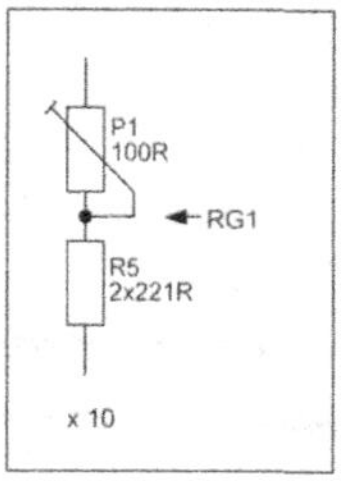

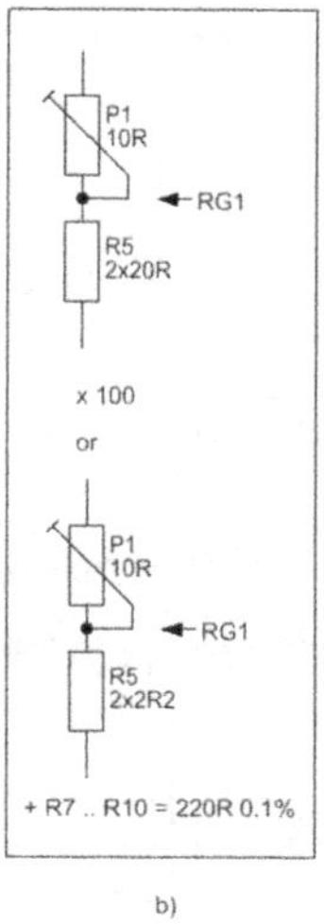

Figure 16 RG alternatives for gains of a) x10 and b) x100 (for only one input stage).

4.4 pmma performance

Table 1 shows a summary of the main measurement vs. calculation results of the MA with gain of x1000. In addition, further down, I show some graphs of the output noise voltage. We see that hum and computer induced spikes will fully disappear if we supply the input with a noise voltage that equals the noise voltage of a resistor ≥ 1kΩ, which will always be the case with DUT output noise voltages ≥ $1\mu V_{rms}$ in B_{20k}.

Item	Unit	ROT	Measured	Calculated
$e_{n.o.tot}$	nV/rtHz	893.0	874.38	871.420
$SN_{o.tot}$	dBV	-78.0	-78.16	-78.189
EIN	dBV	-138.0	-138.16	-138.189
$e_{n.o.1st.tot}$	nV/rtHz			871.188
$SN_{o.2nd}$	dBV		-110.23	-110.926
$SN_{o.3rd}$	dBV		-113.82	-114.687
$e_{n.i.tot}$	pV/rtHz	893.0	874.38	871.420
ein	dBV	-181.0	-181.17	-181.195

Table 1 Gain of 1000 pmma measurement results..

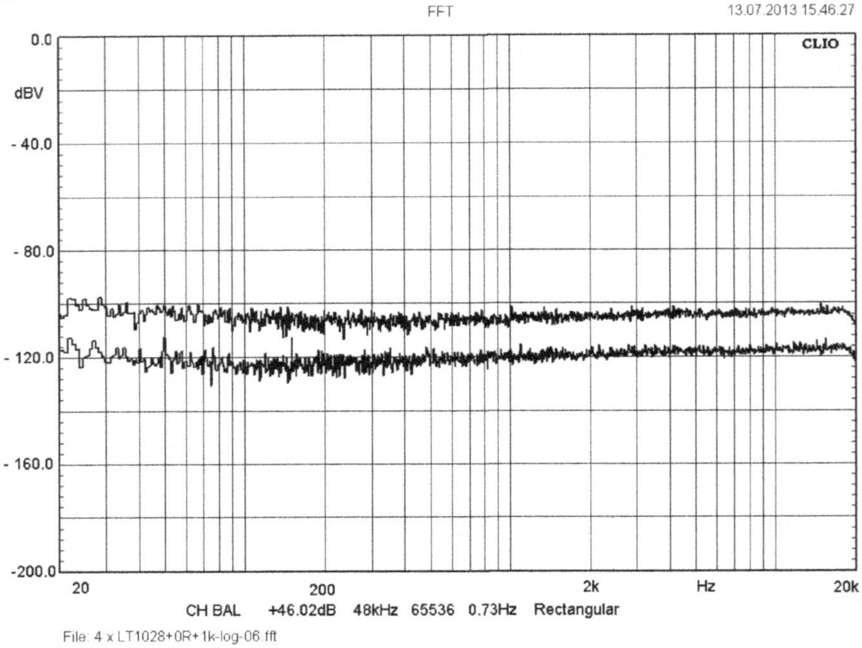

Figure 17 Output noise voltage density curves, including input Cs; lower trace: input shorted; upper trace: input loaded by 1kΩ.

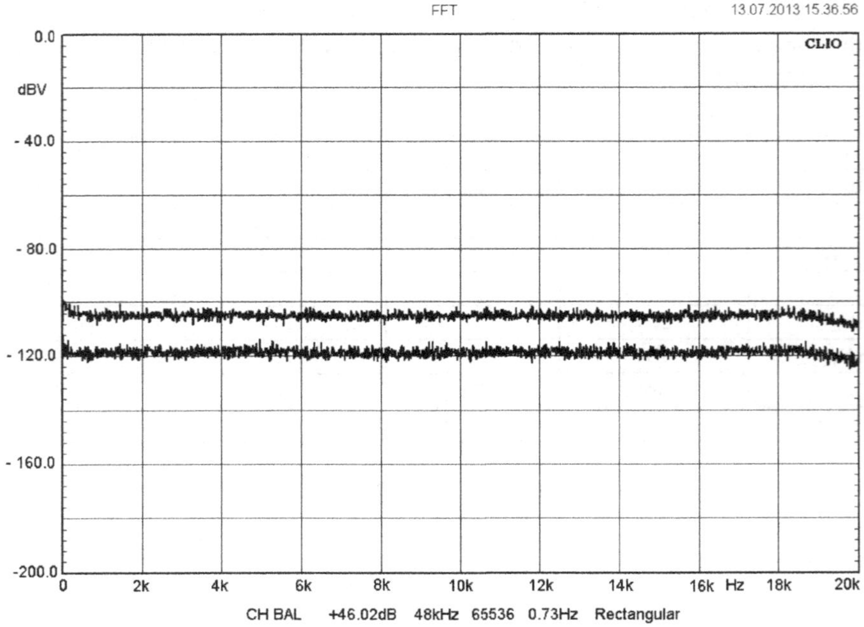

Figure 18 Same as Fig. 17 with linear frequency range.

Notes on **Figs. 17 & 18**:

- On the y-ordinate the gain (+ 40.00dB) of an additional Extra Measurement Amp (EMA; see para 5) and the gain of the measurement filter section (+6.02dB) are taken into account

- FFT resolution is 0.73Hz = sample rate divided by FFT size (length). Averaging in all cases is 50

- No computer-induced spikes could be detected. I could not suppress the tiny 150Hz hum spike that comes in via (my suspicion) the unshielded part of the input XLR plug. It was created by vagabonding hum interferences in my lab; however, with a 1kΩ input load the upper traces show that an input noise voltage density signal of 4nV/rtHz fully hides the spike. The chosen input load simulates a very low-noise amp output (≡ -123.63dBV measured), lower than the original -120dBV goal. Nevertheless, it is really challenging to find the right shielding.

- The heavy lp slope at 20kHz comes from the 6th order 0.1dB Chebyshev measurement lp filter. Together with its hp counterpart at 20Hz, they form a band-pass filter with nearly brick wall character in B_{20k}. The full calculation and the circuit can be studied in [1; Chapters 23 and 26].

Fig. 19 shows the frequency and phase response. At 20kHz, the frequency response is 0.02dB lower than at 1kHz, at 20Hz there is no difference. Taking into account the phase shift of the signal generator's un-bal to bal converter, we find a -5.0°phase at 20kHz vs. -0.25° at 1kHz; at 20Hz the phase becomes 0.15°. The un-bal output shows clipping at 9.9V_{rms} / 1kHz, the bal output at 19.4V_{rms}.

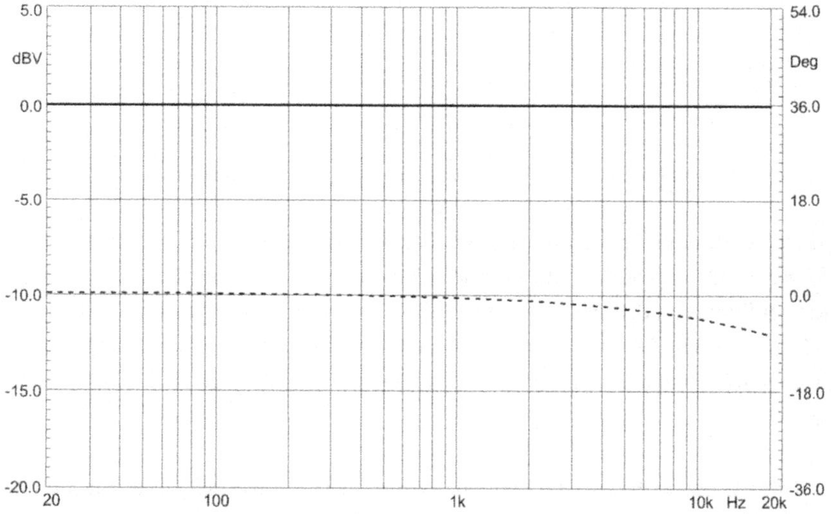

Figure 19 Frequency (top) and phase response (bottom) (bal in / bal out).

5. Practical issues

5.1 Ground loop avoidance

The measurement of extremely low noise levels creates problems in many corners of the project. In most cases, the nastiest ones of these come on board with the connections to the CMS. Balanced lines improve things; however, only a galvanically isolated connection helps avoiding ground loops and the penetration of hum interferences into the measurement set-up. Therefore, I strongly recommend the inclusion of a broadband transformer and associated amplification electronics between MA and CMS.

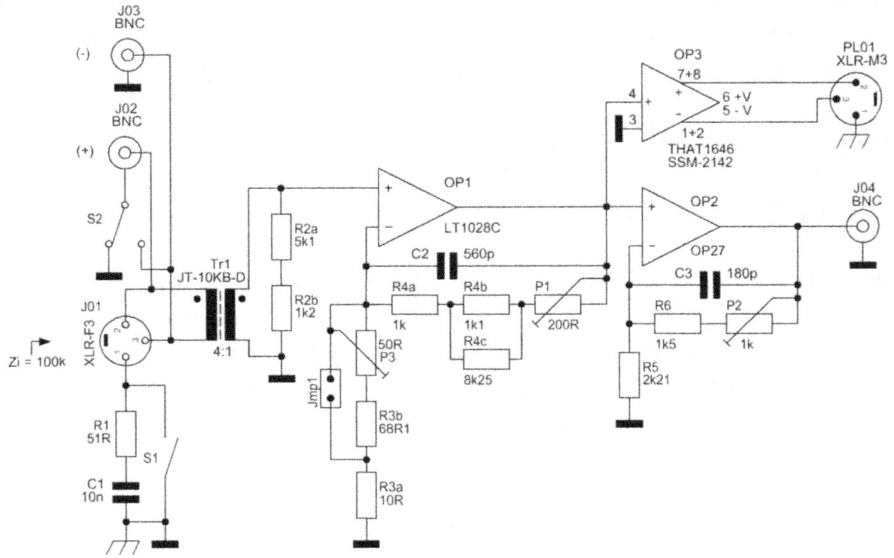

Figure. 20 Transformer driven measurement amp (EMA) for galvanic isolation.

For my own measurement purposes, I have developed such an EMA around the Jensen Transformers 4:1 type JT-10kB-D. In addition to the galvanic isolation it also offers an outstanding CMRR. The electronic configuration consists of an LT1028C op-amp (gains from transformer input to op-amp output x5 - trimmed by P1, or x50 with Jmp1 set and trimmed by P3,), followed by an un-bal to bal and an un-bal to un-bal converter, each with a gain of 2 (SSM2142 or THAT1646 and OP27 plus a trimmed P2).

Fig. 20 shows the details. It has a flat frequency and phase response in B_{20k} and its input noise voltage density is flat in B_{20k}. The measured EMA input referred noise production looks as follow: gain x100: 9.6nV/rtHz; gain x10: 12nV/rtHz. These values are far smaller than the appr. 900nV/rtHz of the preceding MA; hence, there will not be any significant impact on the MA's output noise.

The advantage of the MA - EMA sequence is the fact that the MA can be located very close to the DUT while the EMA might be located far away, closer to the CMS. The balanced lines and the very high CMRR of the EMA insure high immunity against any line induced interference.

5.2 Enclosure

Figs. 21 (right) & 22 show the final MA solution, housed in an Al-box (170x120x55mm^3). In Fig. 21 (left) we can se the Fig. 20 EMA in its case.
At the right lower edge of the pcbs we find the internal psu (±15V), made-up by integrated voltage regulators, fed by a main psu unit that also could feed all other MAs of my collection by a stable ±18V, as long as there won't be any additional ground loop effect. In this case, an separate ±18V ... ±25V psu should be used. The op-amp's PSRRs sufficiently suppress any ripple and noise from the supply lines that might enter the op-amp's internal circuits.

Figure 21 The EMA (left) and the pmma (right).

Figure. 22 The pmma's plug-in input stages.

On the right pcb of **Fig. 22** we can see how I packed Cx and Cy on the feedback resistors.

5.3 Room for improvements

I've already mentioned that the MA has a tendency to run very quickly into overload. In such a case switching the psu off-on will lead to a new stable situation. Further efforts are needed to develop a better behaviour.

The MA's input stages have no support from a DC servo. Therefore, the DC output level slowly moves up and down, roughly ±30mV.

The offset nulling process needs a lot of patience – but once performed we will get at least the result of the previous point. Strong shielding of the MA is essential. The chosen Al boxes work quite well in a rather hum-free environment. This is not the case everywhere, especially if the DUT itself produces heavy vagabonding hum interferences, like eg. valve driven (power) amps. The MA's balanced input allows rather long lines between the DUT and the MA. The MA could thus be placed a distance away - 2m is no problem.

Increasing values of input load resistances lead to increasing differences between the calculated and measured noise production of the MA, the output referred SNs become eg. with $R_{in.rel} = 0\Omega$ or $1k\Omega$ or $2k\Omega$:

=> $SN_o(0\Omega)$ = -78.189dBV calc. and -78.14dBV measured; D = 0.05dB
=> $SN_o(1k\Omega)$ = -64.598dBV calc. and -63.63dBV measured; D = 0.97dB
=> $SN_o(2k\Omega)$ = -61.697dBV calc. and -59.72dBV measured; D = 1.98dB

My interpretation after having checked the data sheet: at the (+) and (-) inputs of each LT1028 the growing imbalance of the source resistances forces its bias current cancellation mechanism to produce additional noise voltage. It has less common-mode behaviour than it would have in the matched case. Therefore, the op-amp's CMRR becomes increasingly less effective. Hence, this additional un-correlated noise voltage sums-up with the data-sheet spec based one. We can study this effect on the two corresponding LT data sheet graphs.

Unfortunately, there is no indication on how to calculate it. Only interpolation of the above shown Ds will lead to calculation results close to the measured ones. MCD-WS-4, Section 4 gives a rather simple derivation of D as function of the input load; it will ease any required interpolation.

6. Recommendations

In a lowest-noise measurement environment, the pmma works quite well, as long as we consider the above-described points. The following paragraphs show how we can use it in general.
We can work through the measurement process and the clarification of potential measurement amp impact according to the following example guidelines:

6.1 DUT output resistance ≤ 10Ω

1	Generally, the determination of the whole measurement arrangement's EIN works by shorting the input, by measurement of the output SN, and by subtraction of the measurement arrangement's gain in dB; here, the pmma's measured input referred rms noise voltage is:	EIN_m	= -138.16dBV
2	Determination of the DUT's output resistance; here, we assume a very low one:	R_o	≤ 10Ω
3	We measure eg. a DUT output referred $SN_{o.m}$:	$SN_{o.m}$	= -130.64dBV
4	We calculate the difference B:	B	= $SN_{o.m}$ - EIN = 7.52dB
5	Fig. 2 shows the impact $W_e(B)$:	$W_e(B)$	= 0.71dB
6	The DUT's real output $SN_{o.r}$ thus becomes:	$SN_{o.r}$	= $SN_{o.m}$ - $W_e(B)$ = -131.35dBV

6.2 DUT output resistance > 10Ω

Depending on the amount of pmma input noise current $i_{n.i}$ any R_o >10Ω will add more or less significant portions of noise voltage, thus, worsening the measurement amp's original EIN to EIN_{corr} by an equivalent amount in dB, hence, we'll get $|EIN_{corr}| < |EIN|$.
Let's check a worst case example with a DUT's total output resistance R_o = 2247Ω, $i_{n.i}$ = 1pA/rtHz, $R_{in.eff} = R_o||R_{in}$ = 2kΩ:

1 Determination of EIN_{corr}: by worsening the original EIN (-138.16dBV) the pmma's total input load $R_{in.eff}$ and input current and voltage noise effect on R_o lead to an additional input noise voltage density $e_{n.i.add}$. With it we can calculate EIN_{corr} as follows:

$$\begin{aligned} e_{n.i.add} &= \sqrt{(i_{n.i}\, R_{in.eff})^2 + e_{n.Rin.eff}^2} && = 5.753\, nV/\sqrt{Hz} \\ \Rightarrow EIN_{add} &= -164.757\, dBV + 43.006\, dB && = -121.751\, dBV \end{aligned} \quad 16)$$

$$\begin{aligned} EIN_{corr} &= 20\log\left(\sqrt{10^{\frac{EIN_{add}}{10}} + 10^{\frac{EIN}{10}}}\right) \\ &= -121.697\, dBV \\ \Rightarrow e_{n.i.corr} &= 5.819\, nV/\sqrt{Hz} \end{aligned} \quad (17)$$

Crosscheck for (17): See Fig. 7 on WS-4. It shows the pmma's output referred $SN_{o.tot}$ versus the total input load R0. We can pick $SN_{o.tot}$ for R0 = 2kΩ and we'll get -61.697dBV. To get EIN_{corr} we have to subtract the pmma gain:
EIN_{corr} = -61.697dBV - 60dB = -121.697dBV

2 We measure eg. a DUT output referred $SN_{o.m}$: $SN_{o.m}$ = -104.38dBV

3 We calculate $G_{loss1.e}$: $G_{loss1.e}$ = 20log[R_{in}/(R_{in} + R_o)]
= 20log(0.89)
= -1.012dB

4 Interpolation term according to Section 5.3/5.: D = 2.0dB

5 We obtain $SN_{o.r}$: $SN_{o.r}$ = $SN_{o.m}$ + $|G_{loss1.e}|$ - D
= -105.37dBV

6 We calculate the difference B: B = $SN_{o.r}$ - EIN_{corr}
= 16.33dB

7 Fig. 2 shows the impact $W_e(B)$: $W_e(B)$ = 0.1dB

8 Without pmma impact the DUT's real output $SN_{o.DUT}$ thus would become:

$$SN_{o.DUT} = SN_{o.r} - W_e(B) = -105.47dBV$$

6.3 Summary of recommendations

Now, we can sum-up all the above given findings into the following main recommendations:

a As long as the DUT output resistance is very low (≤ 10Ω) the pmma measures output referred SNs of DUTs up to -121.8dBV (-129.0dBV) in B_{20k} with an error of ≤ 0.1dB (≤ 0.5dB) only.

b To keep the maximum allowed 0.1dB (0.5dB) border line and by usage of LT1028 op-amp types any DUT output resistances from 0Ω up to 2.25kΩ (equals 2kΩ with the paralleled pmma input resistances) will shift the minimal output referred SNs from measured -121.8dBV (-129.0dBV) to -105.5dBV (-112.7dBV) rounded.

7. Final Notes

- The LT1028 works quite well in an input load environment with nearly equal loads at both inputs. With RGs ≤ 10Ω DUT output resistances with values of RG ± 20% ... 50% do not produce much harm. All cases of larger DUT output resistances require additional calculation efforts, making the calculation results less precise.

- Although handicapped by a doubled input noise current a future project will check the AD797 usability.

- Further CMRR improvement could come from Bruno Putzey's "optimal input biasing", given in Fig. 20 of his very interesting article in L|A Vol. 5. I have to check this too.

- The nasty DC drift at the outputs (TPs 1 ... 4) of the input stages should be controlled by a well-working DC servo. Maybe it could also be the right measure to suppress the tendency of these stages of sometimes running into overload after an input load change. Then, I guess these measures won't prevent the MA from getting the rmma status (rich man's MA).

- Occasionally, in the above given equations we find the term 43.006dB or its rounded version 43dB. It is nothing else but $\sqrt{B_{20k}}$ expressed in [dB] = 20 log(√19,980Hz), and it stands for the difference between the rms voltage of a white noise signal and its equivalent noise voltage density in B_{20k}.

- Of course, the chosen expression "worsening figure $W_e(B)$" is nothing else but another approach to express the noise figure NF_e [dB] of the pmma. However, in this study I use it this way to better underline its significance and negative effect in the measurement process of tiny noise signals

- Finally, here comes the calculation to find the real A-weighted output referred $SN_{o.a.NAD.r}$ = -124.2dBV(A) ± 0.5dB of the NAD M51, without AP analyzer impact, however, in the S-AES17 filter given equivalent noise bandwidth of ENB = B_{20k}, rounded to one digit after the decimal point:

➔$SN_{o.a.NAD.m}$	= -122.0 dBV(A)	= measured A-weighted SN of the NAD M51
	+ 2.0 dB	= A = A-weighting based improvement
=> $SN_{o.NAD.m}$	= -120.0 dBV	= SN incl. 1/f-noise
	- 0.5 dB	= $WF_e(f_c)$ = worsening figure based on the 1/f-noise corner frequency f_c = 350Hz (see LA Vol. 4, p. 19, Fig. 5 or [1], Fig. 7.6a in [dB]!)
=> $SN_{o.NAD.m.wn}$	= -120.5 dBV	= output referred SN excl. 1/f-noise, hence, theoretically white noise only
➔$SN_{i.AP}$	= -122.4 dBV	= AP SYS 2722 bal. i/p referred SN with i/p shorted
=> B	= $SN_{o.NAD.m.wn}$ - $SN_{i.AP}$	
	= -120.5 dBV - (-122.4 dBV)	
	= 1.9 dB	
=> $W_e(B)$	= 2.2 dB	taken from Fig. 2b (see also WS-1)
➔$SN_{o.a.NAD.r}$	= $SN_{o.a.NAD.m}$ - $W_e(B)$	
	= -122.0 dBV(A) - 2.2 dB	
	= -124.2 dBV(A)	

Note: The AP's input referred noise voltage is free of 1/f-noise, hence, if we would calculate B in the A-weighted domain we would create an error that grows for f_c > 1kHz from appr. 0.5dB to > 5dB at 20kHz.

References

[1] 'The Sound of Silence', Burkhard Vogel, 2nd edition, ISBN 978-3-642-19773-4

Experimental Design in Simulation

A Sine/Cosine oscillator - Part 1: Basic theory

Barrie Gilbert

SPICE- the "Simulation Program for Integrated Circuit Emulation" - was developed at U.C. Berkeley in the 1970's by the late Prof. Don Pederson and his students. Today there is an alphabet soup of SPICE derivatives, so the name is used here in its generic sense. For board-level work, one can download a version of SPICE called *MultiSim* (National Instruments) from the Analog Devices web-site. Simulation results presented in this article were obtained using a proprietary simulator developed for integrated-circuit design.

Attitudes toward SPICE range all the way from totally indispensable, through cautiously confident to downright hostile. The feeling is often expressed that "Simulation is all very well: but what *really* happens in practice is invariably quite different!" And this might well be so, as we shall see. Several years ago, I was in a discussion about the power and value of circuit simulation with one of its most spectacularly outspoken critics. He frequently declared that "SPICE lies!" When asked "So, how do *you* do design? Do you use a calculator?" he replied emphatically "Nope! Can't trust 'em! I use green paper and pencil".

It's certainly a matter of choice; but there are many problems with the paper-and-pencil approach to design. First, one needs to use *highly simplified* models for the circuit elements. For example, we allow that a resistor obeys the equation V = IR. But it's not so simple: in a high-accuracy IC simulator, the full model for a resistor requires four terminals and up to twenty-one parameters! Using even the simplest models, we find that networks of moderate complexity, particularly when reactive elements are involved, lead to long and tedious pages of algebra to solve, even allowing approximations. This arduous work might occasionally lead us to important insights. But so will simulation, more quickly and with greater frequency. Further, all active elements (and some passive) are *inherently nonlinear,* and while an idealized paper-and-pencil treatment may be possible and very useful, in certain special cases, others lead to equations that are *mathematically intractable* for one reason or another.

The decision to depend on simulation amounts to no more than what makes the most sense: both from the point of view of achieving a *more accurate* solution for the behaviour of some circuit, and as the best use of one's time. Beyond that, fluency in the use of a simulator often provides unexpected insights into circuit behaviour which might never be reached even by the use of very detailed manual analysis. Accurate simulations are not about trusting the *mathematical engine* which is at their core. Its mundane job is simply to accept an extremely large set of simultaneous equations. When all the "knowns" are inserted it will quickly and efficiently find all the unknowns, such as the voltage on all nodes and the current in all branches. Mistrust stems from an inadequate appreciation of a simulator's unique strengths and weaknesses. It's just a tool, after all; and we have heard that only a bad workman blames his tools.

The *modeling equations* that represent the *behaviour* of one's circuit elements are often too simple or even inaccurate, or both; and the *parameter values* – the *numbers* we must insert into these equations – may also be inaccurate; sometimes they are little more than best guesses. To misquote the old adage, it becomes a matter of "Guesses In, Guesses Out". Accurate models and parameter values are often unavailable to the casual experimenter. Ironically, the more comprehensive the model, the greater is the importance of assigning accurate values to a very large set of parameters.

Even after all this is attended to, in running the actual simulations, the unwary user might not grasp the crucial importance of the *tolerance variables* that control the precision of the solution. Further issues related to the nature of *circuit noise* may not be fully appreciated. In this regard, SPICE can only sum up the *numbers* contributed by individual generators of noise spectral density, from which we can calculate the RMS noise at an output. But crucially, there is no notion in SPICE of noise as a time-varying *stochastic process*.

Precision and Accuracy

We cannot confuse the precision of a solution with its absolute accuracy. If you acquire first-rate device models from a specialist group you are still not yet in a position to undertake reliable simulation experiments until you assign accurate values to numerous parameters. On the other hand, even without accuracy, valuable, *insight-generating* experiments can nevertheless be conducted, relying solely on the *precision* of the models, using the "general knowledge" of the physical world that is built into a modern simulator, including accurate values for numerous fundamental quantities, such as ***q***, the electron charge, ***k***, the Boltzmann constant, π, e, and the like. It can also run circles around us in evaluating integrals and Fourier transforms.

The Sine/Cosine Objective

Any system that oscillates in such a way as to create a pair of state variables representing sine and cosine waves at two separate nodes must invoke a pair of perfect integrals, each with respect to time (dt)

$$\frac{1}{T}\int \sin \omega t = \frac{-1}{\omega T} \cos \omega t \qquad \text{and} \qquad \frac{1}{T}\int \cos \omega t = \frac{1}{\omega T} \sin \tag{1}$$

where ω is the angular frequency and T is the *time-constant of integration*. When a unit-amplitude sinewave is integrated twice, it is multiplied by $-1/(\omega T)^2$. If this output is *inverted* and fed back so as to become the source of the original sinewave, we set up the implicit equation

$$(-1)\text{-}1/(\omega T)^2 \equiv 1 \quad \text{thus,} \quad \omega = 1/T \tag{2}$$

which describes a simple sinusoid. But this idealization says nothing about the *steady-state amplitude* of the oscillation. What *does* determine that in practice? And how do we perform accurate integration in a circuit? Since the earliest days of analog computers, op amps have been used for integration, setting the time-constant T using an external resistor and capacitor (**Figure 1a**).

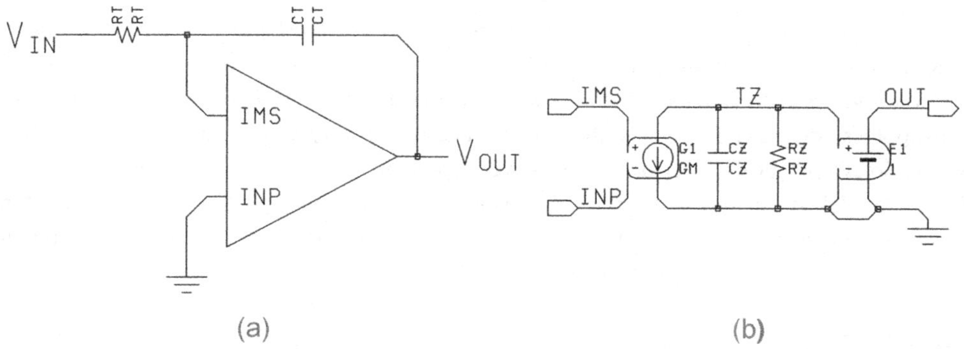

Figure 1 (a) Typical Integrator Connections (b) Simple Model for an Op Amp.

However, this is not a good choice unless T is large (milliseconds), because a modern op amp using dominant-pole HF compensation *already is an integrator!* In fact, an ideal ("text-book") op amp may be modeled as a perfect integrator, having infinite DC gain and a constant roll-off rate of exactly 20 dB/decade in its AC gain. The most important practical aspect of an op amp used as an integrator (and in many other applications) is not its "high DC gain" but rather, the frequency at which its AC gain crosses the unity-gain point.

Figure 1b shows a model for a highly idealized op amp. The input is applied to the differential inputs of a transconductance element, g_m, whose output current is integrated in the "HF compensation capacitor", C_Z, and the resulting voltage is buffered at unity gain before being presented at the output node. In SPICE the g_m element can be realized as a voltage-controlled current-source (VCCS) and the buffer stage as a voltage-controlled voltage-source (VCVS). Although this is an extremely primitive model, it is sufficient for our immediate purposes. R_Z, formed by the internal transistor resistances, models the finite DC gain of this amplifier. The net transimpedance at the TZ node is just the parallel sum of it and the *transcapacitance*, C_Z. In the case shown, with the customary unity-gain buffer, the overall DC gain is simply the product g_mR_Z.

However, few op amps provide a separate ground pin for the TZ elements and R_Z is determined by incremental resistances associated with the transistors, while C_Z is principally their distributed parasitic capacitances, each having a 'back-plate' connected to the power rails (which incidentally results in compromised HF power-supply rejection ratio). Putting these distractions aside, it's evident that the open-loop AC as a function of frequency becomes unity when the product of the g_m and the capacitor's reactance $1//\omega C_Z$ is unity:

$$g_m/\omega_1 C_Z = 1 \quad \text{so} \quad \omega_1 = g_m/C_Z \tag{3}$$

Note that unless R_Z is small, it has little bearing on the unity-gain frequency.

We usually can't control the ω_1 of an off-the-shelf op amp, and we must also note that this frequency is not accurate. It can vary by as much as ±20% over production runs from even a single supplier, due to variations in on-chip resistor and capacitor values; and more when mixing parts from different suppliers. You may have noticed that the variance in ω_1 is very rarely specified in op amp Data Sheets. So as a practical matter, one should never depend on this frequency, called the "gain-bandwidth product". As a matter of fact, this also means that an op amp's -3 dB low-pass corner point for *any closed-loop gain* will be proportionately uncertain. This may be a further contributing factor to the belief that board-level simulations are "unreliable".

The TZ Amplifier

A certain type of op amp differs in important ways from the generic form of Figure 1b. In it, the input g_m cell is replaced with a circuit topology for which the author coined the term "current conveyor" (CC) in the late 1960s, during a period of intense interest in pure current-mode signal-processing – later replaced by a more mature appreciation for what may be called "Free Mode" design [1]. Untold variants of the versatile CC and its applications, largely of academic interest, have been proposed in the decades since.

In contrast to a basic current mirror, which can only accept an input of one polarity and deliver and output of the opposite polarity, a current conveyor can handle *bipolarity* signals at its low-impedance input node, and replicate this current at its high impedance output node, with a gain of unity and high linearity. A further node, at relatively high impedance, determines the potential at the input node; this node is usually, but not necessarily, grounded. A better model and details of its distortion behaviour will be presented in Part 2. For the moment, we can get by with another idealization.

Having the CC cell, and thinking about the possible value of an op amp with a current-mode input pin that acted like a *short circuit* (rather than the usual open circuit) led immediately to the "current feedback" amplifier shown in **Figure 2.** The AD844 is of this form. This product is unique, first, in specifying not only a typical ω_1 but also the guaranteed minimum value; and second, in providing access to the TZ node at a pin. A major advantage of this topology is that the *large-signal response* is determined by the feedback current, directly from the output, rather than from the small, fixed tail current used to bias a differential-pair gm cell. Attention paid to the overall design of the AD844 ensured there will be almost *no slew rate limitation* in its buffer. (It is specified as 2,000 V/μs).

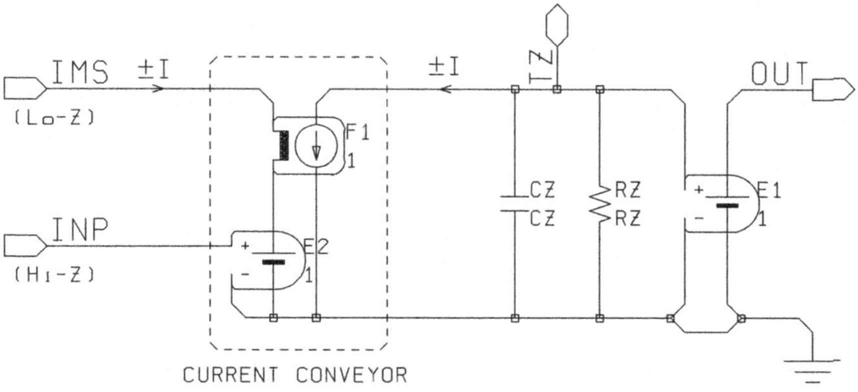

Figure 2 Idealized Topology of the AD844 Transimpedance (TZ) Amplifier.

The "gain" from the current-summing input node to the voltage-mode output is *no longer a simple ratio* (V/V). Instead, it has the dimension of impedance, V/A – the high-valued transimpedance at the TZ node formed by the equivalent internal resistance R_Z (several megohms) in parallel with the small internal parasitic capacitance C_Z, (a few pF), plus optionally, an external capacitor. Consequently, the AD844 can be used to make a high-quality inverting-mode integrator – without any of the complications caused by the "nuisance" poles that arise in a conventional op amp – simply by using an *external* resistor, R_T, from the voltage source to the summing node and adding an external capacitor, C_T, from the TZ node to ground, thus generating a well-defined time-constant of integration $T = C_T R_T$, at its voltage-mode output.

An Ideal Sin/Cos Implementation

Next, we'll examine the essential behaviour of a pair of integrators in a "tail-chasing" loop, using the idealized model of the AD844, as shown symbolically in **Figure 3**. Internally R_Z is specified as typically 3 MΩ, but we will temporarily set it very high, to 10^{12} Ω. (It cannot be omitted from the simulation, since in general the initial potential at the TZ node will not be specified). An ideal unity-gain inverting amplifier completes the circuit for the initial simulations.

Sine/cosine oscillators of this type could optionally employ two different time-constants; and while that will nonetheless generate a quadrature pair of outputs (that is, having a phase difference of exactly 90° - a direct consequence of integration) the amplitudes would be unequal. Therefore it is sensible and customary to make the external resistors and external capacitors equal.

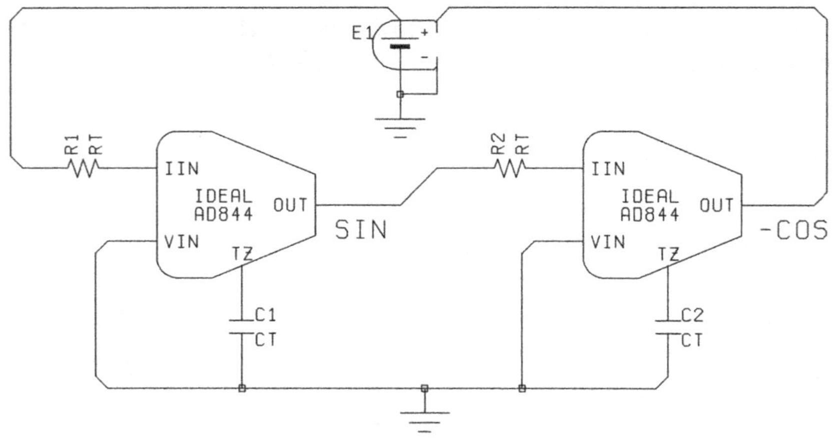

Figure 3 An Idealized Sine/Cosine Oscillator Test-Bench.

The output of the first op amp is labeled 'SIN" and that of the second is "-COS". The purpose of the next few studies is to acquire some basic insights into the behaviour of this simple circuit. In so doing, we will be better prepared to take the next step, using a more complete model for the AD844.

Nowadays, the standard way to enter circuits for study is to use a schematic capture program; and there's little point in constructing the net-list by paper and pencil methods. Of course, in the present case, that would be very simple; but the task quickly becomes far too burdensome in practical design work. Setting $R_T = 10\ k\Omega/2\pi = 1.59155\ k\Omega$ and $C_T = 0.1\ \mu F$, we expect the oscillations to be at exactly $1/(2\pi R_T C_T) = 1.0000$ kHz.

So we press "GO". But nothing happens. Why not?

First, "nothing happens" because there's *no reason* for the condition of the circuit to change. In practice and in the absence of such practical incidents as the sudden application of the supplies (which aren't even included here) it is internal noise that starts things rolling. But SPICE knows nothing about noise as *a random disturbance* to the circuit's state; it can only calculate – blindly and *numerically*, not from physics – how the noise-spectral numbers add up. In fact, the modeling of internal noise as a stochastic phenomenon is possible in advanced simulators; it just requires more computational savvy.

But there's another reason why our circuit "doesn't work". By setting R_Z very high we've generated a pair of poles *right on the imaginary axis* of the s-plane. Thus, there is no reason for an initial disturbance to either *grow* or *diminish*. In practice, even a small amount of excess phase (for example, in the inverter amplifier) may be enough to start the oscillations, and they *will* then grow, until something *pathological* causes the amplitude to *limit*. We need to do better than that. We also want to avoid excess phase in the loop, since this will jeopardize the calibration accuracy at higher frequencies. And importantly, we want to be fully in the driver's seat in the matter of amplitude control.

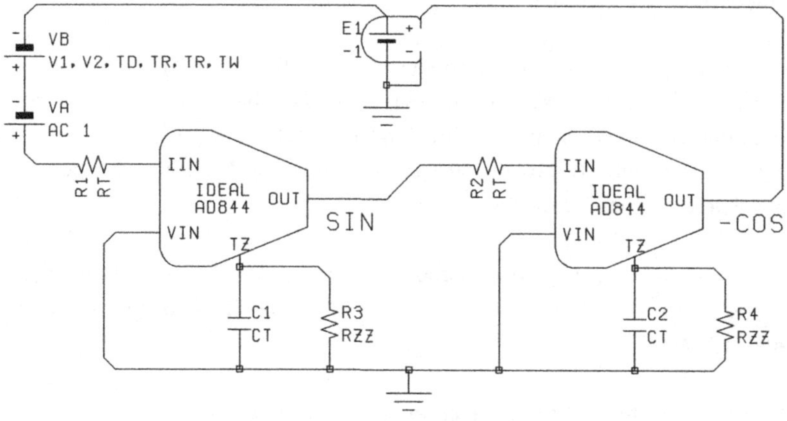

Figure 4 Adding Excitation Sources to the Idealized Oscillator.

So let's inject an AC test signal into this loop as in **Figure 4**. It is provided by the voltage source, VA, in series with R1. Note that an AC stimulus has no "real" amplitude; the assigned magnitude of 1 is only a relative value, allowing us to determine the small-signal (incremental) gain from this input to any other node or branch. Sweeping frequency very close to the expected resonance, we see (**Figure 5**)

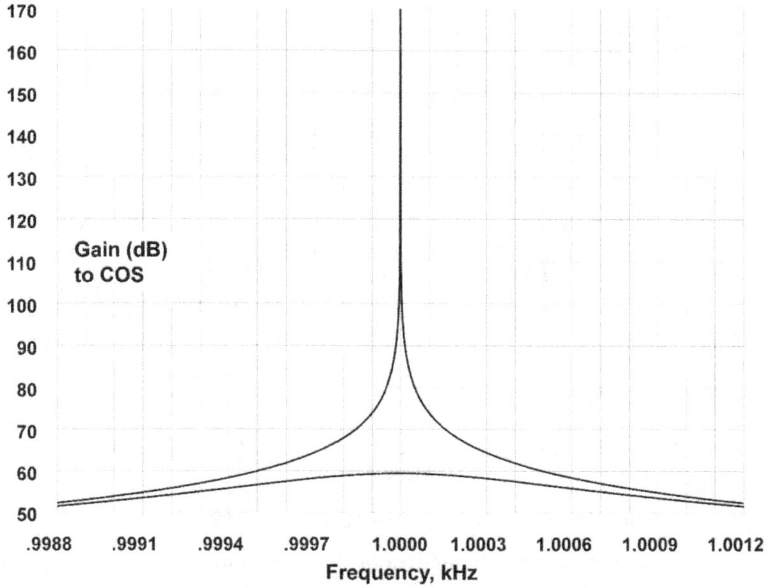

Figure 5 Resonances Observed at 1.000 kHz.

an extremely sharp resonance at 1 kHz with a gain to both sine and cosine output of over 170 dB (316 million). Even with R_Z lowered to 3 MΩ (the typical value for the AD844) the gain is still about 60 dB; but before long, we will cancel the damping of these internal resistors.

Now let's see what happens when we introduce a short excitation pulse by the use of the voltage source VB in Figure 4. In standard SPICE syntax, this is defined by the initial value of voltage, V1; the value it assumes during the pulse duration, V2; the time delay before the pulse starts to ramp, TD; the rise- and fall-times of the ramps, TR; and the 50% width of the pulse, TW. Thus equipped, we can explore numerous different experimental conditions.

Thinking perhaps that all it takes to start oscillations is a short pulse of small amplitude (emulating "a little bit of noise"), try V1 = 0, V2 = 1 mV, TD = 1 μs, TR = 1 ns and TW = 100 ms, setting R_Z now at its internal value of 3 MΩ.

Our oscillator still seems determined to be a damp squib (**Figure 6**). But look more closely, and you will see that the low-level perturbations we've induced are indeed in quadrature and they're *hardly decreasing in amplitude* over time, because having poles that are still lying quite close to the imaginary axis (just slightly into the LH-half-plane because of R_Z) there is very little damping.

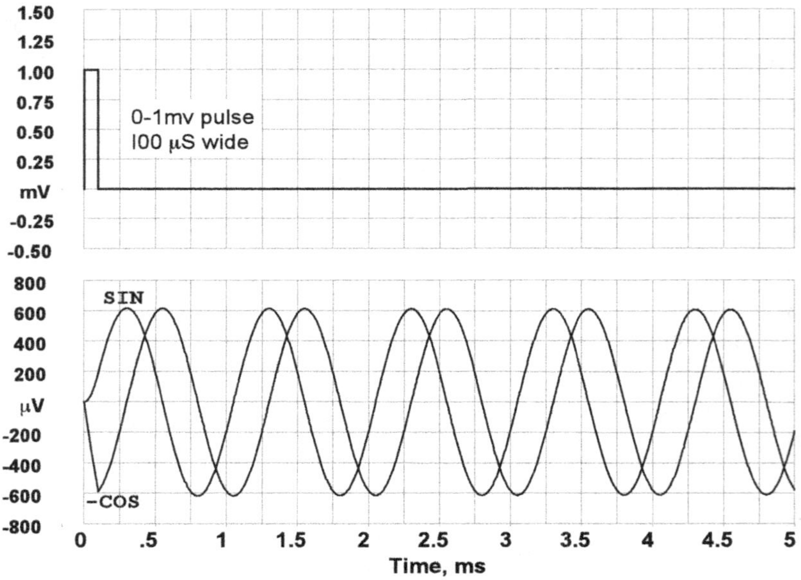

Figure 6 Weak Pulse Excitation of the Oscillator, with R_Z = 3 MΩ

Incidentally, that amplitude of close to 628 μV may strike you as curious (could it possibly be 2πĺ100 μV?) and set you searching for a pencil and notepad. That's interesting, but we mustn't get distracted from making this oscillator work. It is already apparent that the damping caused by any positive

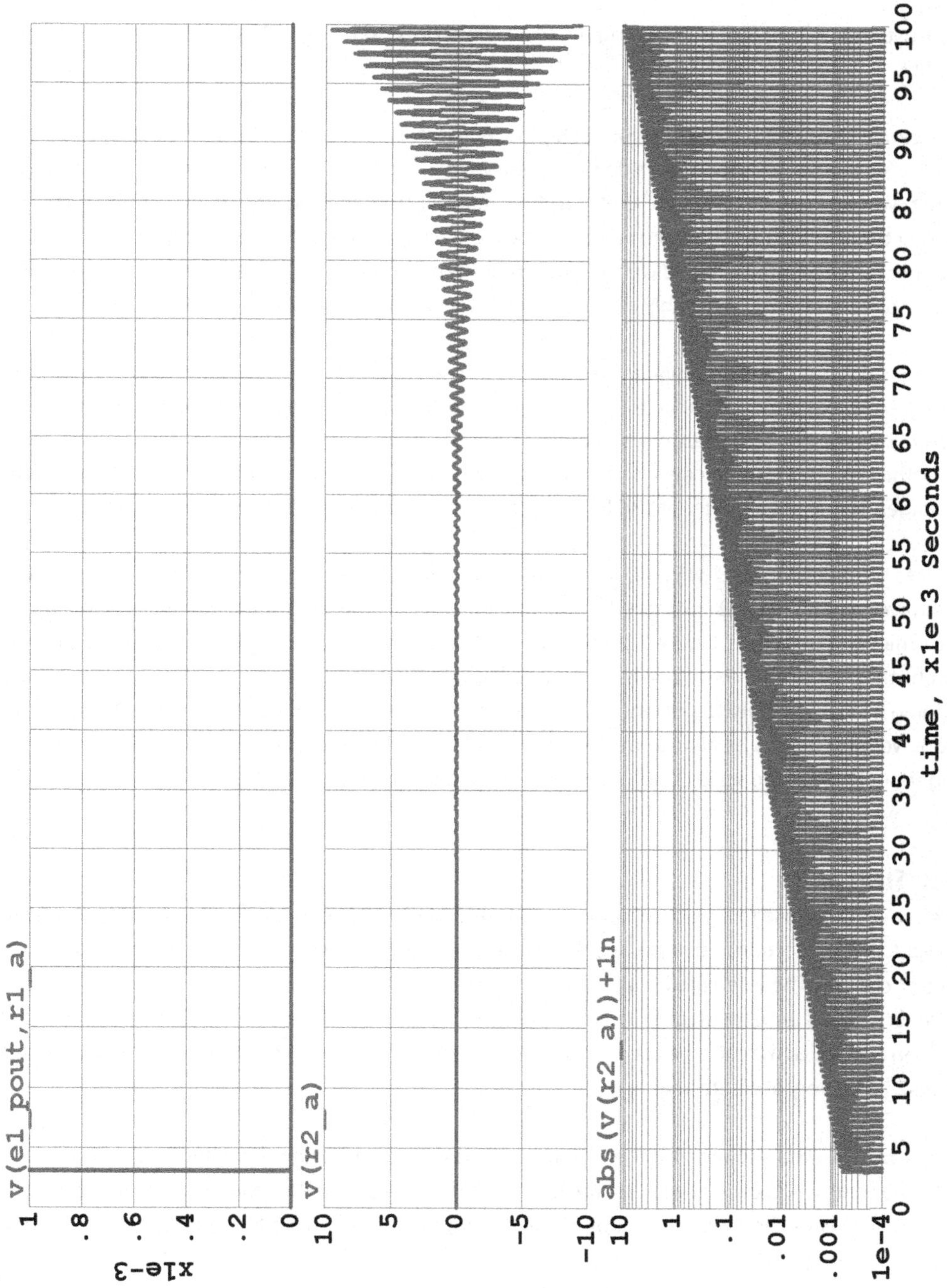

Figure 7 Weak Excitation Pulse of the Oscillator, now with $R_Z = -100\ k\Omega$.

value of R_Z, no matter how large, won't allow it to get airborne and stay aloft; and if you are inclined toward algebra, it's easy to formalize this condition. However, we have gained this *first useful insight* by trusting the simulator to be telling the truth, which indeed it will faithfully do to the fullest of its ability.

The convergence tolerances must be set tightly for these experiments. The most important (and the author's usual defaults for general low-frequency work) are CHGTOL = 1e-20 (that's Coulombs, but don't worry too much about the units for the variables), and ABSTOL = RELTOL = VNTOL = 1p. The tighter these tolerances, the more *precise* will be the results, with less scatter in the data; but of course not necessarily more *accurate!* On the other hand, requiring the SPICE matrix-solver to meet such tight convergence will inevitably slow down the simulation at each *GO*.

We can guess (which isn't forbidden; in fact, *What If?* is a designer's most useful question) that if a positive resistance across the capacitors causes damping, then surely a *negative* R_Z will cause an similar sort of *increase* in amplitude, until "something crashes". Of course, so far in this simulation study, there is nothing *in the circuit* to crash, so before long we will be making megawaves – ridiculously high voltages. This is not a matter of SPICE lying; it's just a matter of appreciating how this can happen.

So let's see what happens when we use similar conditions as before, except that we now set R_Z to the *negative value* of -100 kΩ and start the 1mV pulse at 3ms. (In Part 2, we'll show how negative R_Z can be achieved in practice). **Figure 7** shows that by t = 100ms (that is, 100 cycles) the amplitude is 10V. For further clarity, the SIN output is shown on a logarithmic scale, using the absolute-value function, adding a small quantity to prevent errors as the waves cross zero, which would upset the log function. We must always be alert to the use of *optimal presentation modes* as another insight-generating tool. Here, the log scale clearly shows the exponential growth of the oscillations.

Next Steps

We are getting close to a practical solution. But we'll need much more realism for the AD844 model, to explore its distortion and noise, and thus the spectral quality of the outputs. We may find there are better choices. The interested reader might wish to read the Data Sheets for the Analog Devices AD830, which is another unique amplifier – actually a super-set of the op amp, having very low distortion and which may also be used for accurate integration but in a special connection mode – and the AD734, a very accurate analog multiplier.

Part 2 will show how to monitor the amplitude of a sin/cos pair, using two multipliers to implement a "Pythagorator" – a circuit based on the identity $\sin 2\omega t + \cos 2\omega t \equiv 1$ – and using its output to create a pair of negative resistances of precisely the right amount to keep the poles on the imaginary axis. The Pythagorator is also an integral part of the scheme for varying the amplitude of the oscillations, either manually or electronically. Another pair of multipliers controls the frequency over a wide range, again in either of these two ways. There are some very nice ways to do all these things in very compact circuitry, while holding distortion down to "audiophile" levels.

Part 3 will provide a detailed description of a practical instrument having many uses in the audio lab, complete with measurements to compare against high-accuracy simulation studies.

Reference

[1] B. Gilbert: Current-Mode, Voltage-Mode or Free Mode? A Few Sage Suggestions, *Analog Integrated Circuits and Signal Processing* (Kluwer), February 2004, Vol. 38, pp 83-101

Editor's note: for more thoughts from Mr. Gilbert about analog design and the intelligent use of simulators, see also http://www.youtube.com/watch?v=sN1IN9IM9y4 .

The Prisma Loudspeaker – a shapely adventure

George Ntanavaras

Suppose you want to design a surround loudspeaker for a multichannel system. What are the requirements for such a loudspeaker?
There are many answers to this "simple" question in Floyd Toole's book "Sound Reproduction" [1]. What caught my attention the most was the requirements for the horizontal dispersion as mentioned in chapter 16.5.2 of the book: the surround loudspeaker should have a very well behaved horizontal dispersion out to ±75° at least and ±87° will be even better.
After I finished reading the book, I thought for some time about how to design a surround loudspeaker that would be as close as possible to these requirements. Not an easy task to accomplish.

Usually a surround loudspeaker is mounted very close to the side or back walls of the listening room, so an excellent solution would be a horizontally omni-directional loudspeaker mounted on the surface of the wall or mounted in-wall with its front face flush mounted with the surface of the wall. The in-wall loudspeaker is preferred but in my listening room, and I think in most other listening rooms, it is not possible to place such a loudspeaker; so the only option left is the on-wall loudspeaker.
In this article, I will describe my long-time research for the design of such a loudspeaker and I will present the measurements that I took using "odd" shapes of enclosures since the shape plays a very important role in the polar response of a loudspeaker. Finally, I will present the enclosure I chose for the surround loudspeaker, the procedure I followed to design the crossover and the final measurements of the loudspeaker.

Selecting drivers

First, I had to choose the drivers for the loudspeaker. I wanted to use a full range unit to cover as much as possible of the frequency range to avoid any problems in the polar response that may occur by using two drivers with a crossover. Of course, for the lower frequency range I had to use a woofer. The crossover frequency between the two drivers had to be set to a low frequency, in order to maintain a good polar response
For the full range unit, after a long investigation, I chose the Wavecor FR070WA02 after finding a test report by V. Dickason in the January 2011 issue of the Voice Coil magazine. This is a 2 ¾" unit with a

good on axis response up to 20kHz and a well controlled directivity. The driver is built on an injection-molded polymer frame that is fully vented below the spider mounting shelf for enhanced cooling. The cone assembly consists of a black anodized aluminum cone with a 1" diameter anodized aluminum dust cap. The FR070 has a 22mm diameter voice coil terminated to a pair of gold-plated terminals. Powering this structure is a dual neodymium motor.
The FR070 has its back side opened and it needs a separate enclosure. The Thiele-Small parameters of the unit according to the manufacturer are: Fs=122 Hz, Qts=0.67 and Vas=0.59 liters.

I chose a closed box with a net internal volume of about 0.8 liters. The box resonance frequency was measured at 165 Hz with a Quality factor of Qtb=0.85. The same volume filled with absorbing material was used more or less for all the enclosures for the Wavecor FR070 that I constructed. For the woofer, I select the Seas ER18RNX-H1456-08 unit. I had used this driver for other projects with excellent results. It is a 6.5" reed/paper pulp cone driver with a 12 mm peak-to-peak linear excursion and up to 22 mm maximum excursion. According to the manufacturer data the ER18 woofer has the following Thiele-Small parameters: Fs=37 Hz, Qts=0.32 and Vas=32 liters.
I chose a closed box with a net internal volume of 6 liters. This gave a rather high box resonance frequency at 90 Hz but with a very welcome Quality factor of Qtb=0.7. Since I already had a subwoofer in my system this was not a big issue. The same volume was used as a basis for all the enclosures for the ER18 that I constructed.

A note about the measurements

I used an EMM-8 microphone with a MP-1r preamplifier from iSEMcon, Germany. According to the manufacturer, the microphone was calibrated against a reference microphone Bruel & Kjaer 4133 and I loaded the calibration file into the ARTA software that I used for the measurements and the analysis [2]. So any possible errors from the microphone response were compensated for.
It was not possible to do the measurements outdoors so I did them indoors in my large listening room and this introduced errors. To avoid these errors as much as possible, I placed the loudspeaker in the middle of the room with its back on the floor and its front side looking at the ceiling of the room. Since the loudspeaker will be mounted on the wall, the above placement simulates quite well the real mounting of the loudspeaker; a very welcome feature. The microphone was placed at a distance of 0.6 meter for some of the enclosure types and at 0.8 meter for the others.

The above setup gave an almost reflection free measurement of about 6 msec. Although this is a rather good result for in room measurements, it does not have the accuracy and the resolution of an anechoic room measurement.
For each type of enclosure I performed a typical set of measurements consisting of the following:

a. The on axis response measurement;
b The measurement at 30 and 60 degrees horizontally to the right of the axis. These are noted as H+30 and H+60 in the graphs. Because of the symmetry of all the enclosures in the horizontal

plane I did not measure them at 30 and 60 degrees to the left side;

c. The measurement at 30 and 60 degrees below and above the loudspeaker axis. These are noted as V+30 and V+60 for the measurements above axis and V-30 and V-60 for the measurements below axis.

A little theory

When a loudspeaker is mounted on a wall it radiates only in 2π or half space. This means that the radiation that would have traveled into the rear hemisphere of the loudspeaker is reflected forward. In other words, there is a reflected acoustical image of the source. Depending on the distance between the real and the image source, an acoustical destructive interference will occur at some frequency. This frequency will not be constant but it will depend on the angle of the measurement point in reference to the main axis. As the measurement point moves further off the axis the distance between the real and the image source is reduced and the destructive interference frequency is increased. **Figure 1** shows an example. Loudspeaker L1, which is similar to the type S1 that I will describe later, is mounted on the wall. An image of the loudspeaker shown as L2 is created.

Now let's suppose that a measurement microphone is placed in position P1 on the main axis of the loudspeaker at a distance of 0.8 meters. The center of the acoustical source of the loudspeaker is placed in front of the driver at the point C1 and has a distance of 12 cm from the wall. The corresponding

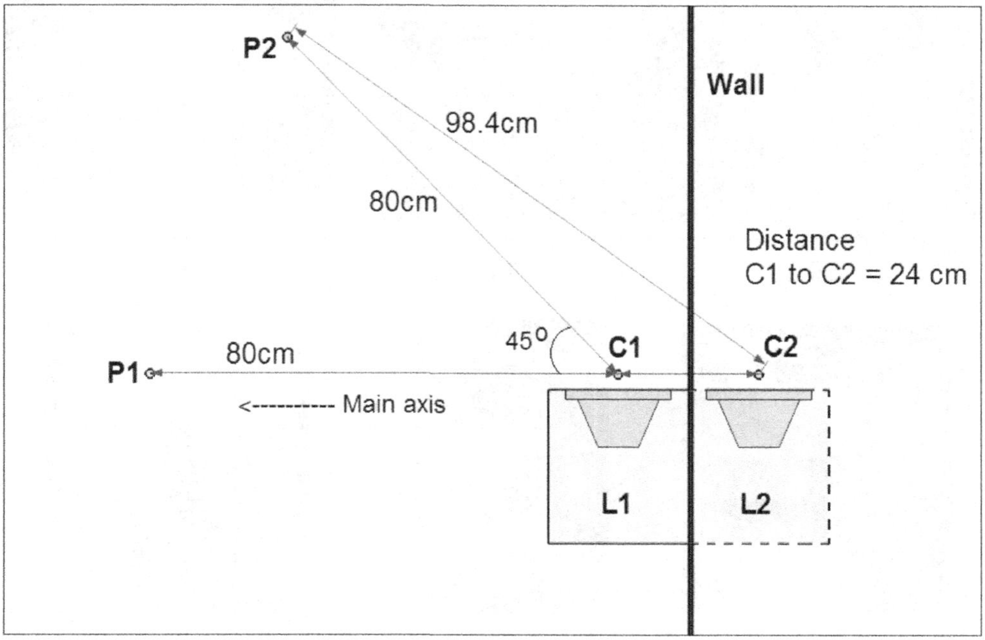

Figure 1 - On wall mounting of a loudspeaker.

center of the image loudspeaker at point C2 will be at a distance of 80+24 = 104 cm from position P1. When the loudspeaker radiates a sound with a wave length which is $\lambda/2=0.24$ meters the image loudspeaker will radiate the same sound out of phase at point P1 and an acoustical dip will occur. The frequency will be: $f=c/\lambda = 344 / 0.48 = 717$ Hz where c is the speed of sound in meters per second.
Now let's move the measurement point at P2 which is 45 deg above the main axis and at the same distance (0.8 meters) as before from the center of the acoustical source. The distance of the center of the image loudspeaker from point P2 is now 0.984 m so the acoustical destruction will occur when $\lambda/2=0.984-0.80=0.184$m or at a frequency $f=c/\lambda = 344 / 0.368 = 935$ Hz.
The directivity of the driver will affect the above results but it is clear that the placement of the loudspeaker near the wall will create a broad dip in the power response of the loudspeaker. And as stated at page 192 of the above mentioned book of Floyd Toole, this does change the sound of the loudspeaker. How much, it depends on the acoustic characteristics of the room in which the loudspeaker radiates.

Photo 1 - The enclosures investigated and presented in this article.

Looking for the proper enclosure

I investigated a lot of different enclosure shapes during my research and I build and measured most of them. I used foam board with hot glue to simplify the construction and reduce the material cost for each enclosure. This is a very light material, not recommended for loudspeaker enclosure construction. To overcome the problems coming from the reduced stiffness of the enclosure, I used several internal braces from the same material and a lot of sound absorbing material inside the enclosure. Also I cut off all frequencies below 150 Hz from the test noise signal to reduce the low frequencies vibration of the enclosures. All these measures worked well for this project and I had acceptable results. Most importantly, the time and the cost for modifying and building the next enclosure was the minimum possible.

From all the enclosures that I have investigated and measured, I will describe the 6 most interesting of them as shown for reference in **photo 1**. To keep the presentation as simple as possible, I named these enclosures S1, S2, S3, S4, S5 and S6.

Enclosure S1

As I mentioned, my initial plan for the loudspeaker was to use a closed box woofer to cover the frequency range up to about 400 to 500 Hz and a full range unit for the higher frequencies.

I mounted the woofer on the top side of the enclosure for two reasons; first to minimize the distance between its acoustic center and the wall and second to provide a uniform horizontal polar response since the main listening axis of the loudspeaker will be perpendicular to the axis of the woofer. I decided to use two full range units for a wider horizontal polar response. I placed them on the front of the woofer box with 60 degrees angle between their main axes to cover uniformly the horizontal plane.

I placed the microphone 0.6 meters in front of the mid-point between the two full range units which I considered the main axis of the loudspeaker.

The top of **figure 2** shows the horizontal polar response of the woofer direct driven without any filter. The dip in the response due to the wall proximity starts at about 750 Hz for the on axis measurement and goes up to 1 kHz when the measurement is done at 60 degrees off axis. Otherwise the horizontal polar response is uniform and independent of the measurement angle due to axial symmetry of the radiator in the horizontal plane.

The vertical polar response is shown on the bottom of figure 2. It remains uniform and independent of the measurement angle for frequencies up to 300 Hz. Above this frequency, the woofer becomes directional and the response begins to deviate up to 15 dB.

Similarly, **figure 3** shows the polar response measurements for the full range unit. The distance of the acoustic center of the full range unit is estimated to be about 0.27 meters from the wall so the dip in the frequency response due to wall proximity is now in the range 250 Hz to 300 Hz depending on the angle of measurement.

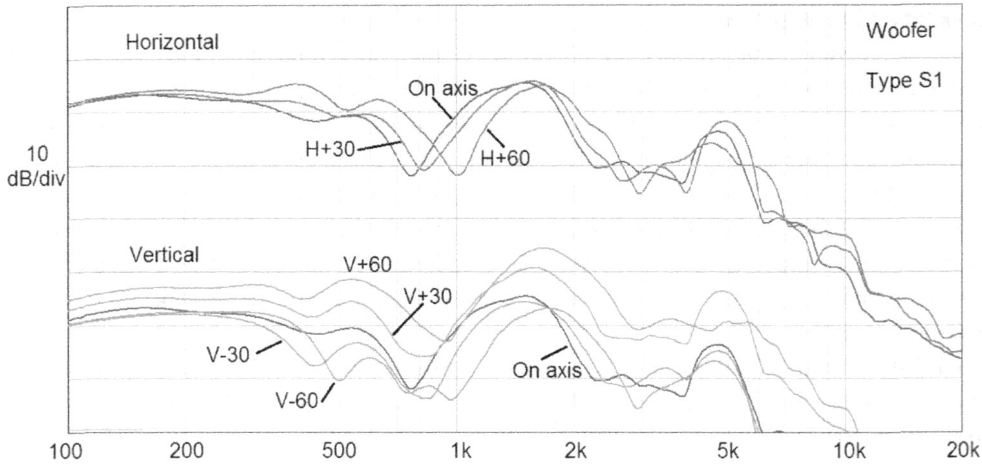

Figure 2 - Type S1, Woofer polar response

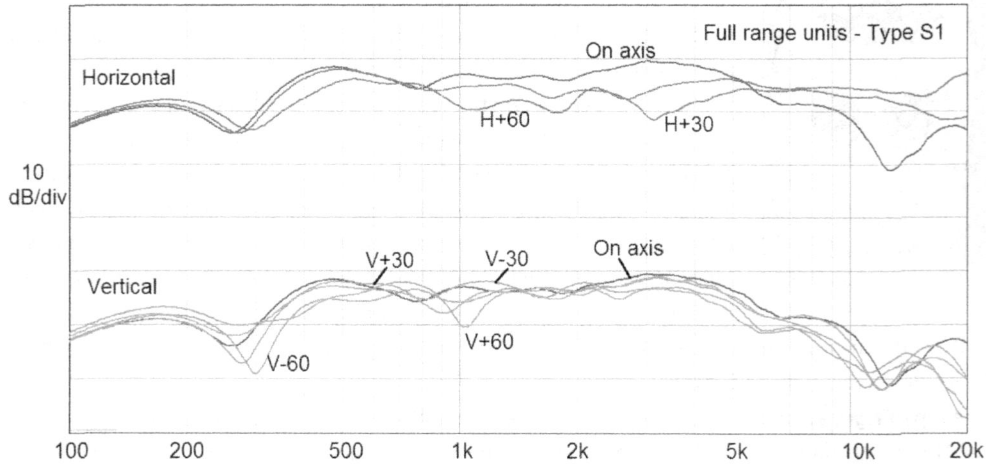

Figure 3 - Type S1, Full range speaker polar response

The horizontal polar response is very uniform up to about 900 Hz. Above this, it starts to deviate depending on the measurement angle. When the measurement axis is only 30 degrees off axis the maximum deviation reaches 10 dB in the range around 3 KHz. This is mainly due to the destructive interference between the two full range units because of their finite distance as was proved by disconnecting one unit and measuring again. As expected, the response is better at 60 degrees off the main axis of the loudspeaker because at this angle, the measurement axis coincides with the axis of the single full range unit.

The vertical polar response is very uniform due to the symmetry of the drivers, but above 4 kHz begins to fall as expected since the full range drivers are placed at 30 degrees off the main axis.

I was not satisfied with the above results of the measurements so I decided to continue looking for a better solution and following a step by step procedure, changing one parameter at a time.

Enclosure S2

I started by looking for a solution to the interference problems of the two full range units that existed in the previous measurements. I thought that by increasing their distance, the interference will move to a lower frequency out of the operating range of the driver. I constructed the enclosure S2 to examine the above. I placed the two full range units to the left and the right side of the enclosure.

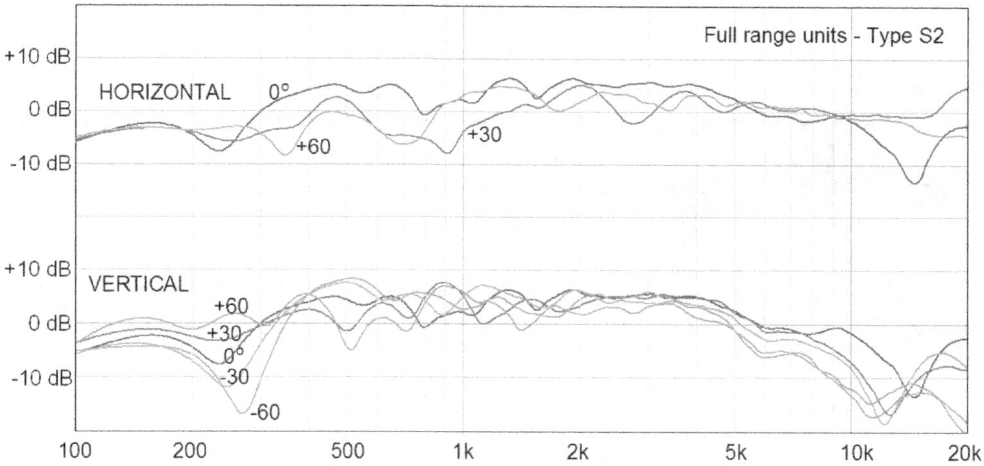

Figure 4 - Type S2, Full range speaker polar response

Figure 4 shows the result of the measurements for the full range unit. As expected the range of the frequencies with the destructive interference has indeed moved lower, in the range between 600 Hz to 1 kHz. Above 1.5 kHz the horizontal polar response of the full range units is very well behaved. The vertical polar response is also very well behaved above 400 Hz, because of the symmetry of the drivers.

The polar response of the woofer is shown in **figure 5** and as expected has a lot of similarities with the polar response of the Type S1 woofer. The dip in the response due to proximity to the wall remains at the range between 700 and 1.1 kHz.

I continued with a new modification for the full range drivers.

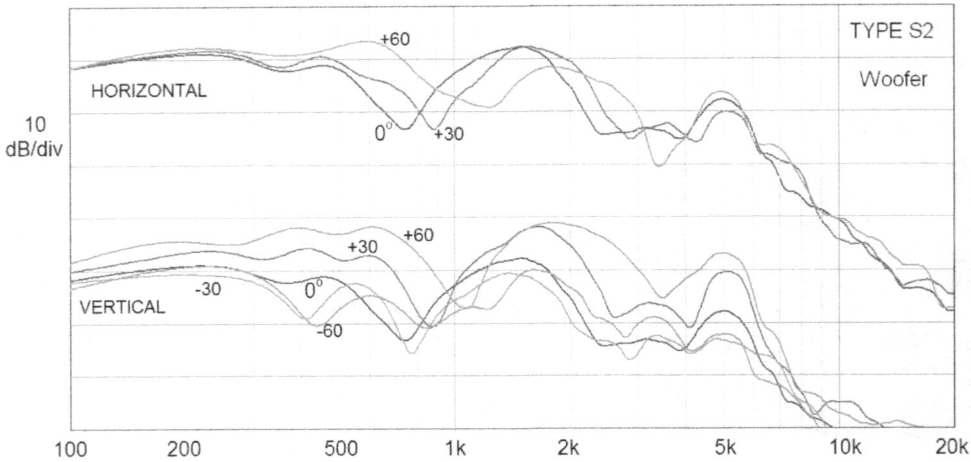

Figure 5 - Type S2, Woofer polar response.

Enclosure S3

In this enclosure, I placed the full range units close to the back side of the enclosure to minimize their distance from the wall and also to maximize the distance between the two units. The position of the woofer remained in the same as before.

The polar response of the woofer is shown in **figure 6** and it more or less similar with the polar response of the previous enclosures.

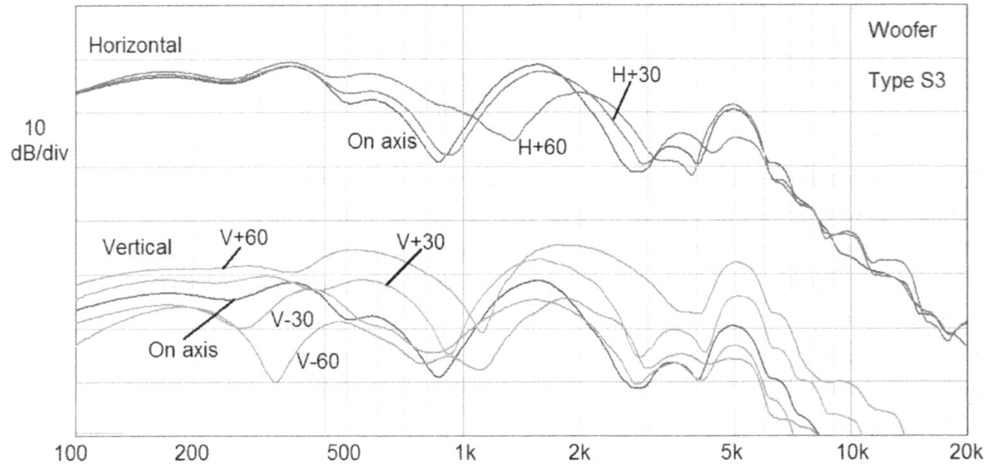

Figure 6 - Type S3, Woofer polar response.

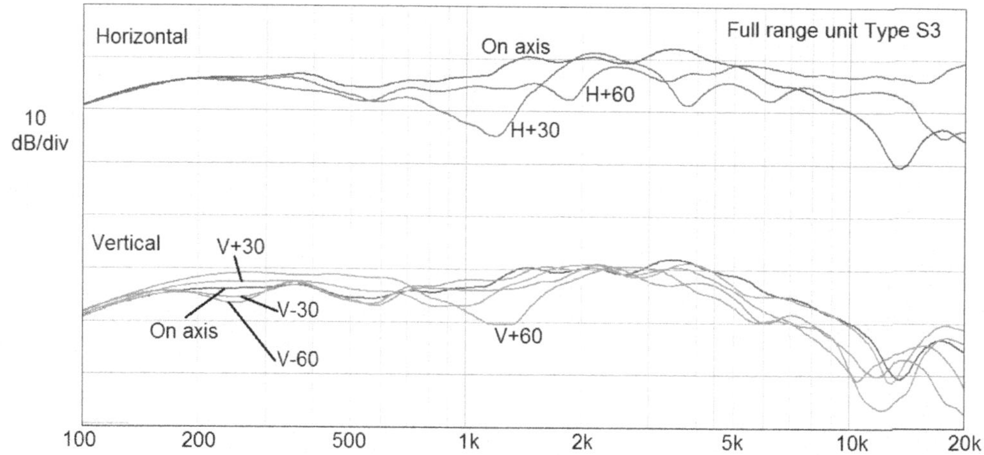

Figure 7 - Type S3, Full range speaker polar response.

Figure 7 shows the measurements for the full range unit. With the exception of a dip at 1.2 kHz when the measurement axis is at 60 degrees above, the vertical polar response is good up to about 4 to 5 kHz. Above 5 kHz the response falls because the full range drivers are placed at 30 degrees off the main axis.

The horizontal polar response is very well behaved up to about 700 Hz. Then, it starts to deviate depending on the measurement angle. At 30 degrees off axis, the maximum deviation reaches 10 dB in the range around 1.3 kHz. Again the response is better at 60 degrees off the main axis of the loudspeaker because it is on axis for the single full range unit.

Enclosure S4

After the polar response measurements of the types S1, S2 and S3 and the problems of the destructive interference, I decided to continue the investigation by using only one full range unit. The woofer remained in the same position because I thought that maybe it would be possible to have a good polar response with the single full range unit down to 300 Hz, below which the woofer could take over. I placed the full range unit in the front side of the enclosure to be as far as possible from the wall, leading to enclosure S4.

The polar response of the woofer is shown in **figure 8**. The horizontal polar response is quite good up to 700 Hz. Above this frequency, the usual dips in the response exist due to the proximity of the wall. As expected, the vertical polar response varies a lot above 200 Hz due to the inherent directivity of the woofer.

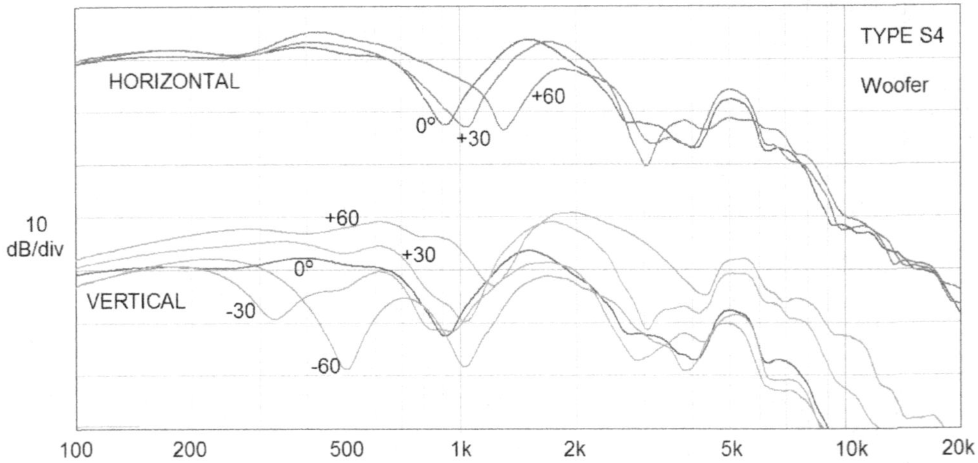

Figure 8 - Type S4, Woofer polar response

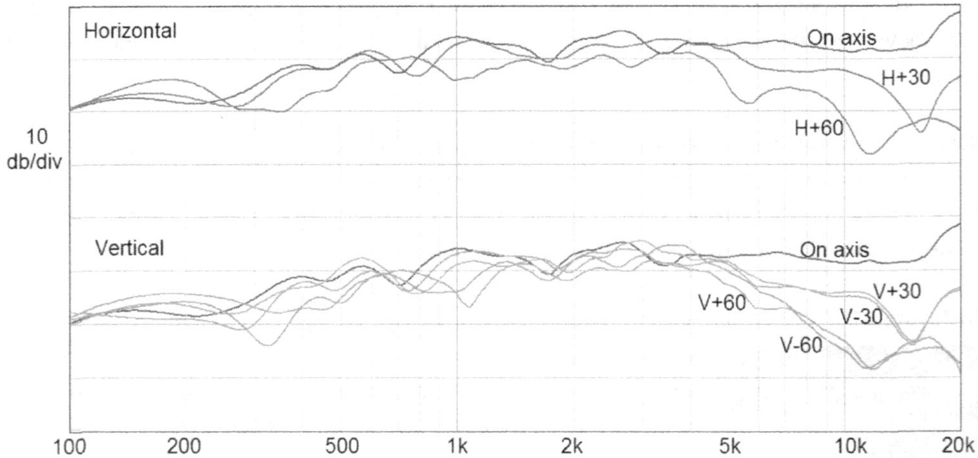

Figure 9 - Type S4, Full range speaker polar response.

Figure 9 shows that the measurements are much better now for the full range unit. Both the horizontal and the vertical polar response have very well behavior up to about 4 to 5 kHz. Above 4.5 kHz the unit becomes directional and the response falls smoothly.

The next step was to find a similar better polar response of the woofer.

Enclosure S5

All the previous enclosures had a similar woofer placement. The variation was at the mounting of the full range unit(s) in different positions in the enclosure to study the various issues arising during the research.
With the type S5, I decided to investigate a different woofer mounting while keeping the full range unit in a position that gave good results in the previous measurements.
I mounted the woofer on the top side of the enclosure in a manner that was not perpendicular to the wall but angled. The full range unit was mounted just below the woofer in the front side of the enclosure and far from the wall.

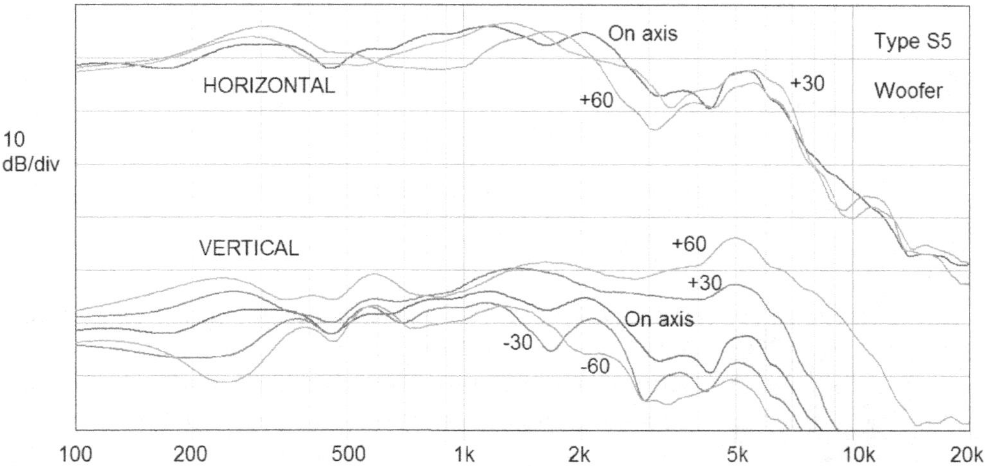

Figure 10 - Type S5, Woofer polar response.

Figure 10 shows the response of the woofer direct driven. The horizontal response is very well behaved and the dips in the response due to the proximity of the wall were eliminated. Only a very small dip around 800 to 1 kHz exists when the measurement is performed 60 deg off axis.
The frequency response in the vertical plane is dependent on the measurement angle although in the range between 400 and 1.2 kHz is much better. Above this frequency, the response varies considerably due to the inherent directivity of the driver.

Figure 11 shows the measurements for the full range unit. Both the horizontal and vertical response are very well behaved. Above 4.5 kHz the response falls as normally is expected from the directivity of a 2″ unit.

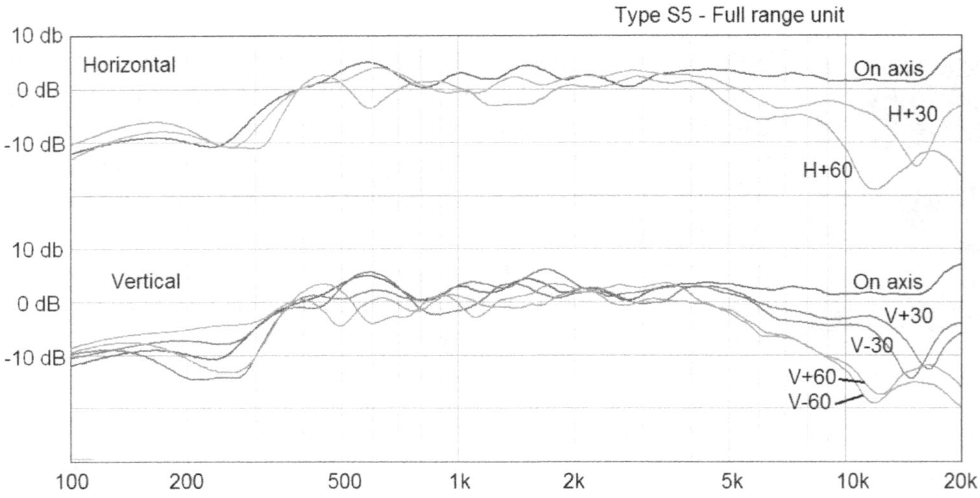

Figure 11 - Type S5, Full range speaker polar response.

Enclosure S6

Until now, I had built and measured enclosures with "strange" shapes. As I was thinking how to improve the measurement of the woofer's polar response, I decided to build a more classic rectangular enclosure to compare its response with the previous results. In all previous measurements, the response of the full range unit began to fall over 4.5 kHz, so I decided to solve also this issue by making a three way system. This was something I had wanted to avoid from the beginning because it will complicate the design, but after all these measurements it appears that it is not possible to have a good behavior in the polar response over the full frequency range by using a two way system.

I kept the same Seas woofer and the Wavecor FR070WA02 full range unit was used only as a midrange unit. For the high frequencies I decided to add the TW022WA04 tweeter also from Wavecor. This is a 22 mm design with 4Ω impedance. It uses a neodymium magnet for lower resonance frequency (750 Hz) and distortion. Its recommended operating range is from 2 kHz to 40 kHz.

I wanted to minimize the depth of the enclosure and thus the dip in the response due to the proximity to the wall, but the depth of the basket and the magnet of the woofer did not permit the total depth of the enclosure to be less than 10 cm. I was thinking of finding another woofer with less magnet depth when I got a better idea. Why not try to fool the sound waves into thinking that the distance from the wall is less than it actually is.

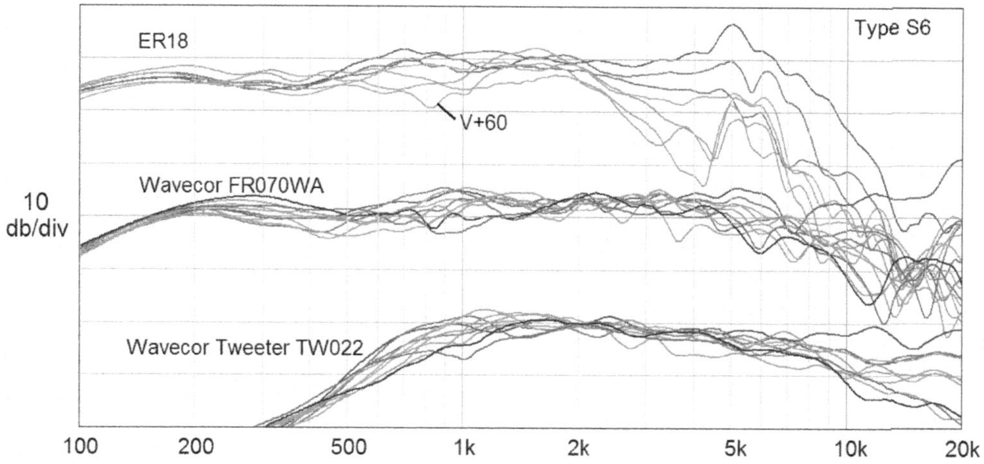

Figure 12 - Type S6, Polar response of the drivers.

To achieve this, I extended the front side of the enclosure to the back side with beveled sides so that it ended smoothly to the wall without the 90 degrees steep slopes of the sides of a rectangular box. The result is the type S6 enclosure.

Figure 12 shows the very well behaved polar response of all the drive units for the S6 enclosure. It is so uniform that there is no need to show separately the horizontal and the vertical response as in the previous examples.

The polar response of the woofer remains quite uniform up to about 2 kHz. No dip worth mentioning exists in the response due to the proximity of the wall because the front side of the enclosure finishes smoothly to the back wall. This was confirmed by removing the beveled sides of the enclosure. A small dip in the frequency range around 800 Hz exists only when the measurement is performed at 60 degrees above axis.

The midrange polar response is also very uniform up to 4.5 kHz. Above this frequency, the response falls smoothly as expected for a 2" unit. Again no dip due to the wall proximity is present in the measurements.

The tweeter polar response is also excellent up to about 8 kHz. Above this frequency the response falls smoothly as the 22 mm diaphragm becomes more directional.

The final Prisma enclosure

If you have studied carefully the previous measurements, you can guess that I chose the Type S6 enclosure for the surround loudspeaker because it gave the smoothest polar response for all the drivers. For obvious reasons, I named it "Prisma".

If I had to make a second choice, it would have to be the type S5 enclosure, probably after doing some further modifications to improve the woofer response.

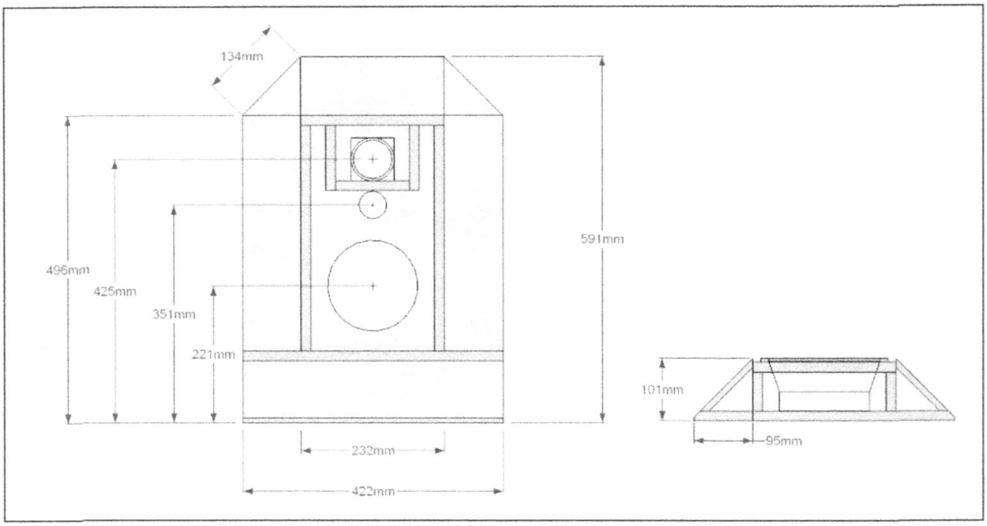

Figure 13 - The Prisma enclosure drawing.

Figure 13 shows the drawing of the Prisma enclosure; the height is 591 mm, the width is 422 mm and the depth is only 100 mm. It is larger than I would prefer, but these dimensions are needed to give the necessary net internal volume of 8 litters and also to mount the crossover.

The enclosure was constructed from 20 mm MDF. I covered the inside of the front panel with a 50-50 mix of sand and wood glue as shown in **photo 2**. As we will see later, this did a good job for damping any possible vibration on the large front panel.

I filled the box rather tightly with long fiber wool to attenuate any internal resonance of the box.

The Wavecor midrange was mounted in an internal separate enclosure with a total volume of about 1 liter. The net volume is about 0.6 liters and was fully stuffed with long fiber wool to attenuate any internal resonances.

At the bottom of the enclosure an additional space with access from the back side was created for the mounting of the crossover.

I painted the whole enclosure with a wood imitation varnish and this offered a nice look to the loudspeaker.

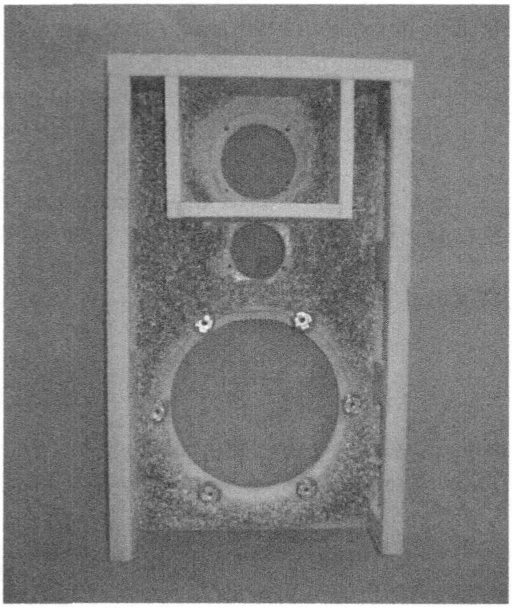

Photo 2 - The inside of the front panel showing the damping material.

Photo 3 - The completed Prisma loudspeaker without the front grille..

Photo 3 shows the completed loudspeaker without the front baffle.

Designing the crossover

With the construction of the enclosure finished, my attention now turned to the design of the crossover. I mounted the driver units and I measured again their polar response. The results were very similar with the type S6 shown in figure 12 and I will not repeat them here.

From these results I chose the crossover frequencies; 500 Hz between the woofer and the midrange and 3 kHz between the midrange and the tweeter. I also decided to use third order Butterworth alignments for all the filters. This alignment gives a flat on axis response and a flat power response.

I started the designing of the crossover with a DSP based active version. I measured the on and off-axis response and did the necessary modifications. When I was satisfied with the measurements, I listened to the loudspeaker using well known program material. Additional modifications were required before I was satisfied with the final active version, which I then used as a target to design the passive crossover.

My first step was to equalize each driver of the loudspeaker, for a reasonably flat on axis frequency response well beyond their operating range. I used my Behringer Ultra-drive Pro DCX2496 as the equalizer. From its outputs I drove each unit separately with 3 power amplifiers.

I avoided using the equalization capabilities of the DCX2496 like time delays or very narrow notches that will be difficult to implement with a passive crossover. Fortunately the response of each driver was uniform enough as shown in figure 12 and no complicated filters were required.

After that I entered the crossover topologies and frequencies for each driver set into the DCX as well as the relative level for each driver.

I measured the response of the complete system and made any necessary adjustments. When I was satisfied with the measurements I started listening to the loudspeaker using well known program material. The listening was done with the loudspeakers in a stereo configuration. Further adjustments were required during the listening mainly at the reproduction levels of each driver. When I was satisfied, I saved the configuration file to the memory of the DCX2496. Then I continued by setting another crossover topology and equalization of the drivers and the new configuration file was stored to the DCX.

In this way, I could easily restore all these different configuration files for comparisons by listening to very well known music tracks.

After several tries with different crossover alignments and a lot of listening tests, I arrived at the following settings: the crossover frequency between the woofer and the midrange was set to 520 Hz. A third order low pass filter was used for the woofer and a band pass third order filter for the midrange. The crossover frequency between the midrange and the tweeter was set to 2.5 kHz. After trying both 3rd and 4th order filters, I decided to use the higher slope 24 dB/octave of the 4th order filter mainly because the polar response was much better and the higher attenuation protects the tweeter better from the high excursions at the lower frequencies.
With the design of the active crossover finished, my effort now turned to the design of the passive crossover.

I used the LspCAD 6 demo version for the design of the passive crossover [4]. The demo version contains all the functionality of the unlimited version with a few exceptions: it is not possible to save the projects; it is not possible to export data and the crossovers are limited to two driver units.
Although the program has a lot of capabilities, I used only its capability to optimize the components of a crossover filter so that its response will be close enough to a given target response. This target response was entered to the program as an ASCII text file. I did the optimization for only one driver unit each time.
I measured the response with the active crossover at the input of each driver and I set it as the target function for the design of the passive network. The program also needs the impedance of each driver when mounted in the enclosure which I measured and entered as a text file.
It is very-very important to select a proper topology for the filter in order for the optimizer to work efficiently. For example, it is not possible to enter a second order topology and to expect the program to optimize it for a given target slope of 20 dB/octave. When all the necessary data were entered and the optimizer started, it was pure magic to watch the current filter response to change as the program tried to match the target response. Although an exact match between the response of an active crossover based on a DSP and the response of a passive crossover cannot be achieved, it was close enough when I stopped the optimizer.

I used a piece of wood to quickly assembly the prototype, I connected the speaker units and I measured the frequency response of each part of the crossover.
Only a few modifications were necessary to the crossover parts to match the simulation results with the real measurements. The value of the resistors R3 and R7 which adjust the tweeter level had to be decreased and also the resistors R4, R5, R6 and R8 have to be added in parallel to some of the capacitors and inductors to suppress resonances in the filters response.

The final crossover of the Prisma loudspeaker is shown in **figure 14**.
The components L2, L3, L8, C1, C4, R1 and R4 form the third order filter for the woofer, which actually has a 20 dB/octave slope. The components C1, L8 and R1 are used to compensate the peak in the impedance of the woofer at 95 Hz due to the box resonance. This was necessary because otherwise it was not possible to approximate the target response at the frequencies below 200 Hz.

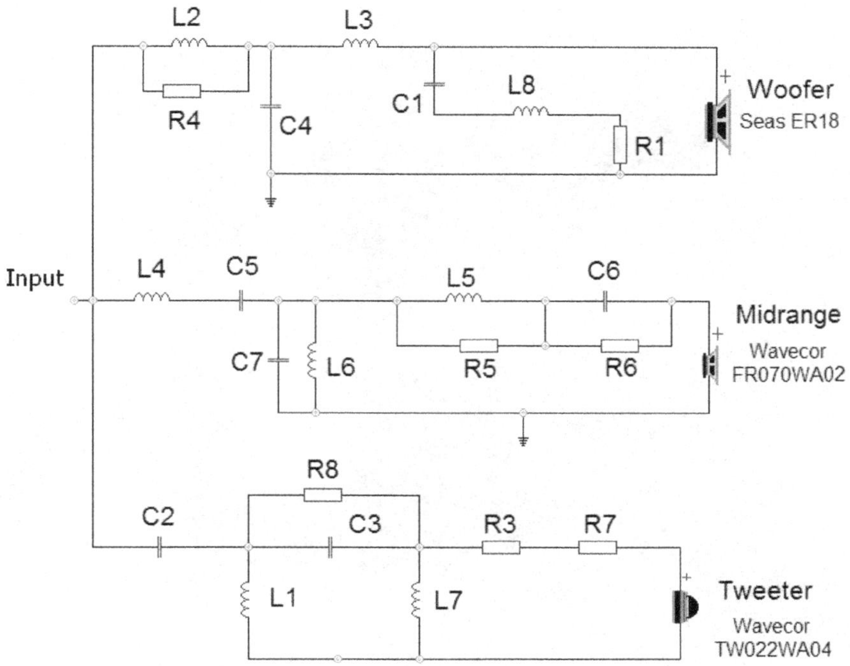

Figure 14 - The Crossover for the Prisma loudspeaker.

The components L4, L5, L6, C5, C6, C7, R5 and R6 form the 3rd order band pass filter for the midrange. The components L1, L7, C2, C3 and R8 form the 4th order high pass filter for the tweeter. The resistors R3 and R7 attenuate the tweeter level.

For the final assembly of the crossover, I glued all the components with hot glue on a piece of plywood which had the form of the base of the enclosure as shown in **photo 4**. Two loudspeaker terminals for banana plugs were used as input connectors for the crossover. I placed and glued the crossover in the bottom of the loudspeaker enclosure. **Table 1** lists the parts used for the final passive cross-over.

Resistors	Inductors	Capacitors
	(All non-ferrite core, except as noted)	(All 400 V, MKP, except as noted)
R1=6.8 Ω / 10 W	L1=0.39 mH, 0.42 Ω	C1=150 uF / 100V, electrolytic
R3=4.7 Ω / 10 W	L2=4.5 mH, 0.4 Ω	C2=5.6 uF
R4=100 Ω / 5 W	L3=1.5 mH, 0.23 Ω	C3=8.2 uF
R5=15 Ω / 5W	L4=1.6 mH, 0.3 Ω	C4=68 uF
R6=15 Ω / 5W	L5=0.25 mH, 0.18 Ω, Air core	C5=12 uF
R7=3.9 Ω / 10 W	L6=1.8 mH, 0.33 Ω	C6=56 uF
R8=82 Ω / 5 W	L7=1.2 mH, 0.23 Ω	C7=12 uF
	L8=18 mH, 4 Ω	

Table 1

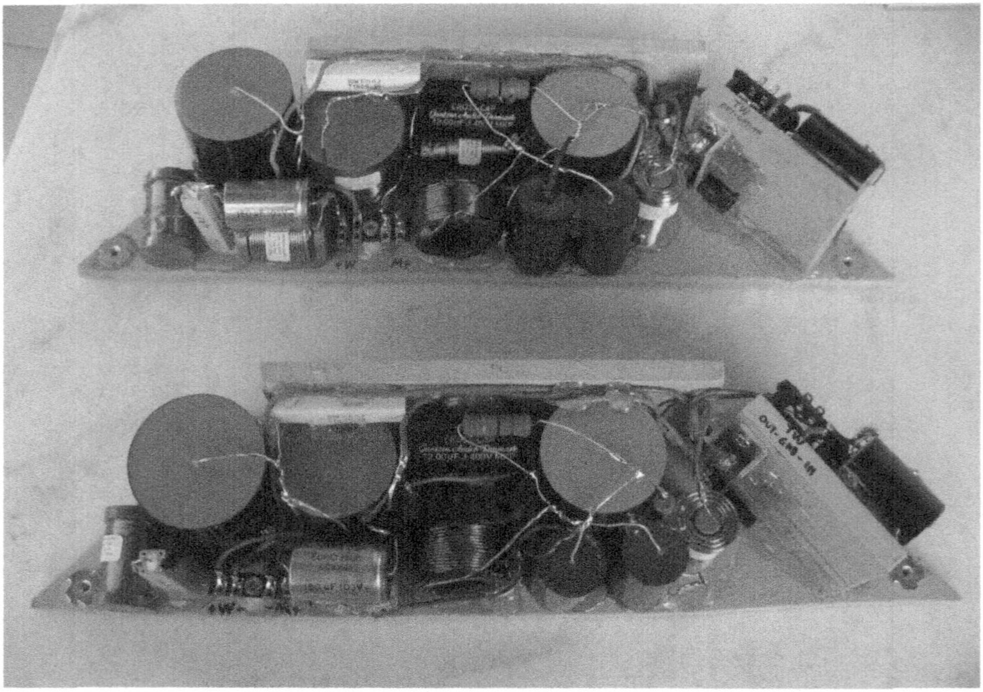

Photo 4 -The two assembled crossovers boards for the loudspeakers.

The final measurements for the Prisma loudspeaker.

When I completed the loudspeaker, I took a lot of measurements to evaluate its performance. I will present some of them below.

One of the targets of the crossover design, not mentioned until now, was to keep the impedance of the loudspeaker greater than 5 Ω. This is very important because usually the length of the cable between the surround loudspeaker and the amplifier is rather long and a low loudspeaker impedance, in combination with the cable resistance and the amplifier output impedance, could cause frequency response irregularities,.

The loudspeaker's impedance magnitude and phase is plotted in **figure 15**. The minimum impedance of 5.3 Ω occurs at around 4.3 kHz but the phase is only -4 degrees. The other minimum of 7.5 Ω occurs at 1.2 kHz with the phase also only -5 degrees. The impedance peak value is 13 Ω and occurs at about 2.1 kHz. The phase of the impedance ranges from a minimum value of -30 degrees and a maximum value of +25 degrees.

The horizontal polar response of the loudspeaker is examined in **figure 16**. The measurements were taken on the axis of the loudspeaker, at 30 degrees, at 45 degrees and at 60 degrees. The responses

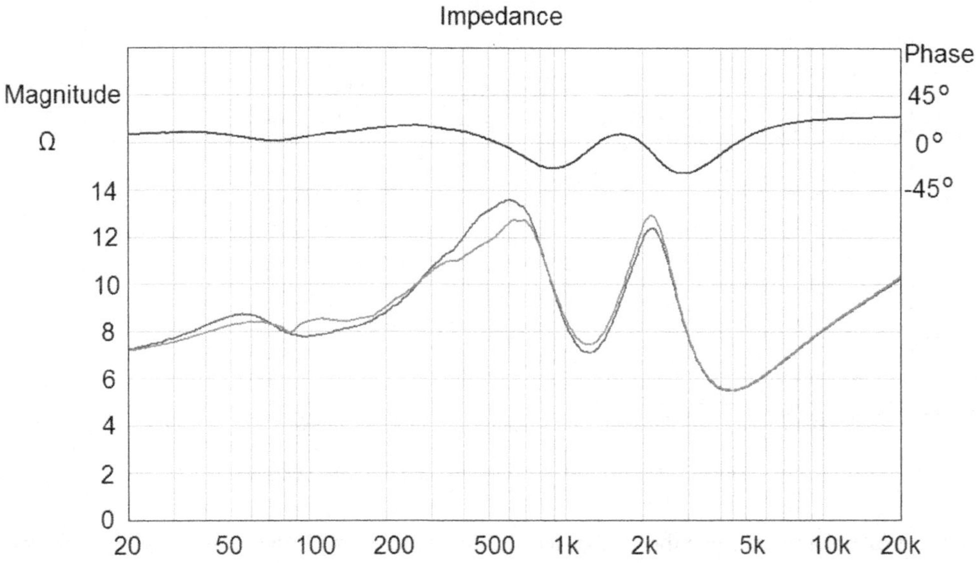

Figure 15 - The impedance of the Prisma loudspeaker.

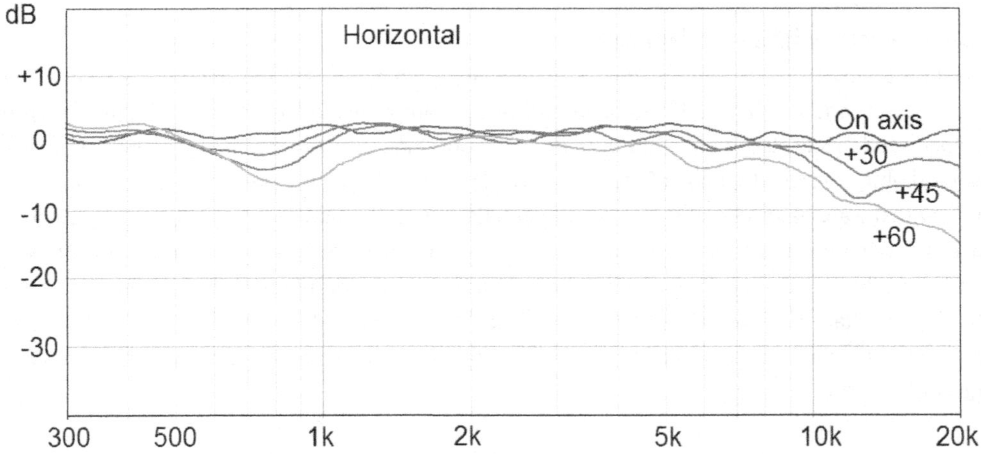

Figure 16 - The Horizontal polar response of the Prisma loudspeaker

are very good except in the range 700 to 1000 Hz where a small dip of 5 to 6 dB is present when the measurement is performed 60 degrees off axis.

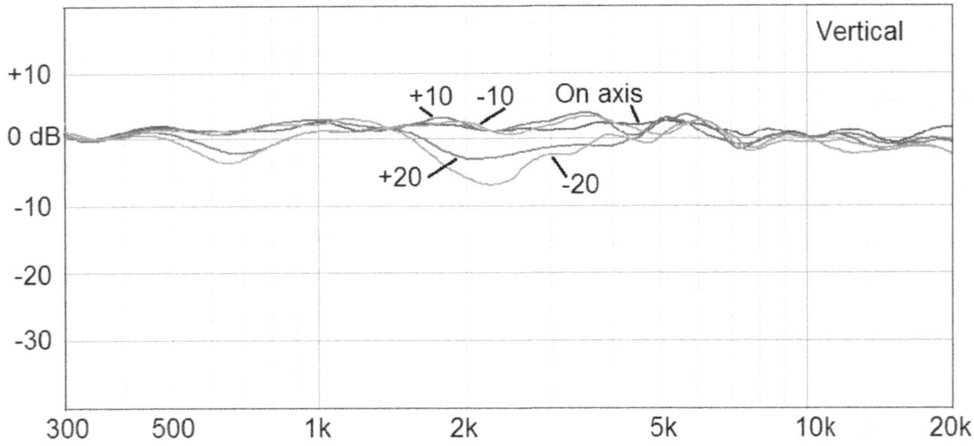

Figure 17 - The Vertical polar response of the Prisma loudspeaker

The vertical polar response of the loudspeaker is shown in **figure 17**. Here the measurements were taken on the axis of the loudspeaker, at 10 degrees and at 20 degrees above and below the main axis. The responses are very good except in the range of 600 to 700 Hz where a small dip of about 3 dB is present and in the range around 2 - 2.5 kHz where another dip of 5 to 6 dB is present when the measurement is performed 20 degrees below axis.

Measurements with an accelerometer

Next I used my accelerometer system to test the Prisma loudspeaker. I designed this system a few years ago. It consists of an ACH-01-03 accelerometer, an amplifier / processor that also computes the velocity and the displacement of the acceleration input signal and an external 12 VAC power supply. It is a slightly improved version of the design that I presented in audioXpress magazine [5, 6].

First with the accelerometer system I investigated the panel vibration behavior of the enclosure. The accelerometer was attached to the panel under investigation using a very thin double sided tape. I had to shield the ACH accelerometer for these measurements, because a strong interference of the 50 Hz line voltage was present at all measurements. This was easily done with a 5 cm X 5 cm piece of aluminum foil around the accelerometer, which then was connected electrically to the amplifier ground.

As an excitation signal I used pink noise produced by the ARTA software. I mounted the woofer on the front panel and a power amplifier drove it at high sound level.

I used an additional gain of 20 dB on the accelerometer preamplifier because the signals measured by the accelerometers were very low in level. The additional gain was then compensated by the "Scale" function of the ARTA software by -20 dB down.

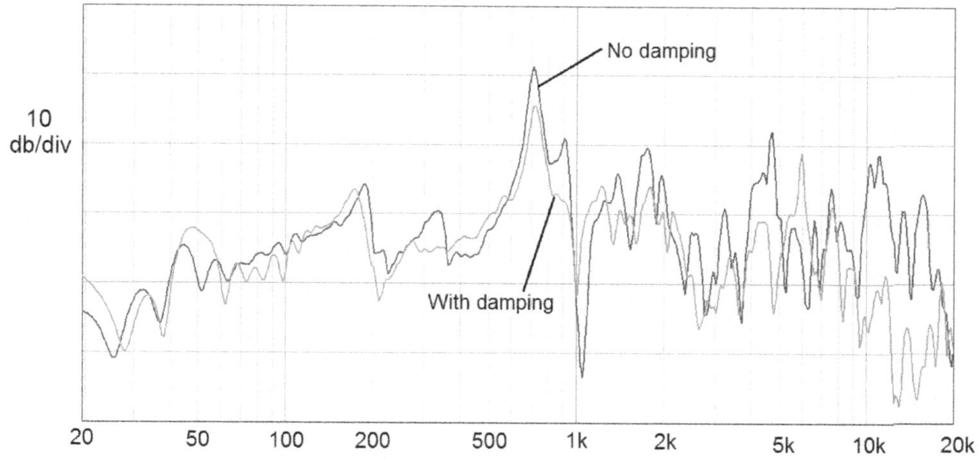

Figure 18 - Accelerometer measurements for the effectiveness of the dampening of the front panel.

Figure 18 shows an example of the measurements that I took. The top curve is the acceleration of the front panel when no damping material is used. The bottom curve is the same panel when the damping material was used consisting of a 50-50 mix of wood glue and sand that covered the inside of the front panel as shown in photo 2. The damping performs well in the frequencies above 150 Hz. It reduces the peaks at 330 Hz, 913 Hz, 1.4k, 1.95k, 2.5k, 3.2k, 4.7k and in the range above 10 kHz.

I used the accelerometer system to perform some other interesting measurements concerning the voice coil displacement (Xmax) of the woofer cone in relation to the applied voltage drive and the Total Harmonic Distortion (THD) of the reproduction.
These measurements were made possible with the STEPS part of the ARTA software which offers this very powerful capability. For more information see [3].
I used a very thin double sided tape to attach the accelerometer to the center of the woofer cone. The accelerometer measures the cone acceleration and after processing the accelerometer signal, the accelerometer amplifier output provides the voice coil's displacement.

ARTA's STEPS program increases the drive voltage to the woofer in steps at the selected frequency and measures the T.H.D. of the woofer sound pressure output with the microphone and the displacement of the woofer voice coil with the accelerometer. I measured the cone displacement at several frequencies to test the woofer capabilities. **Figure 19** shows the measurement for the 30-Hz frequency. The woofer needs 8 Vrms for a peak displacement of 2.8 mm with a 10.8 % distortion. When the drive voltage rises to 10 Vrms, the peak displacement reaches 3.3 mm but now the distortion is 13%. **Figure 20** shows the 60-Hz measurement where things start to improve. The woofer is driven with 18 Vrms and produces a peak displacement of 6 mm with a distortion of only 8.5%.
The above measurements provide interesting insight on the woofer capabilities to produce low frequencies at large peak displacements.

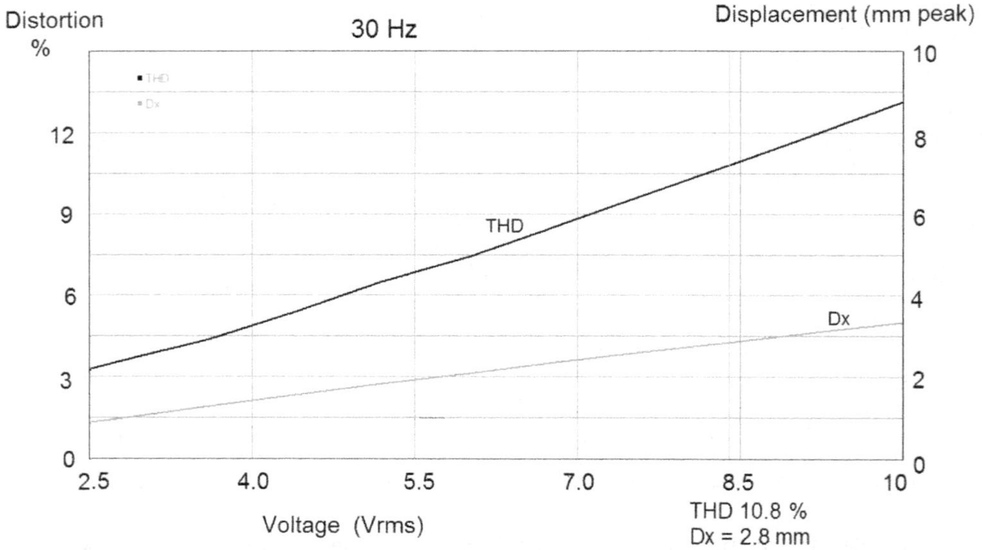

Figure 19 - The woofer cone displacement and distortion at 30 Hz

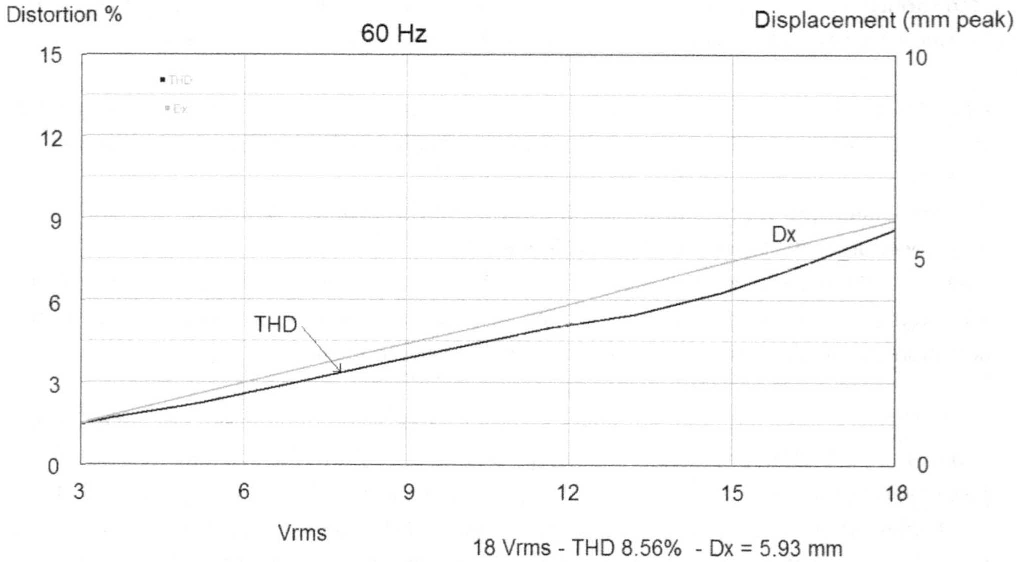

Figure 20 - The woofer cone displacement and distortion at 60 Hz.

Conclusion

The Prisma loudspeaker was a hard project. I spent most of my free time and a lot of effort designing, measuring, and finally acoustically adjusting the loudspeaker. But I think the result was worth the effort. The sound of the Prisma loudspeaker is really good and I am very pleased with its performance. Although it was designed for the surround channels of my multi-channel system, it can also be used as the main loudspeaker of a system as long as it is mounted on a wall and a subwoofer is added.

References

[1] Floyd Toole: "Sound reproduction, Loudspeakers and Rooms", Focal Press, 2008

[2] Measured with the demo version of ARTA software; http://www.artalabs.hr/

[3] http://www.artalabs.hr/AppNotes/AP7-EstimationOfLineardisplacement-IEC62458-EngRev01.pdf

[4] LspCAD program: http://www.ijdata.com/LspCAD_demo.html

[5] G. Ntanavaras, "Accelerometer testing of Loudspeaker drivers", audioXpress, June 2011.

[6] G. Ntanavaras, "The Horizon subwoofer", audioXpress, September 2013.

Author's note: I have available some high quality PCBs (with solder mask and silk screen) and engraved Front Panels for the construction of the accelerometer amplifier described in [5]. Interested parties should email me at: gntanavaras@gmail.com.

Component-level mutual induction

Stan Curtis

This month I'd like to tell you a story of how I spent nearly 40 years trying to solve an intractable problem only to find that I cannot fully explain the workings of my solution. And so we have a tale of two parts; my quest for solutions and a call for help in explaining the outcome. It all starts with the effect the layout and construction of a piece of electronics can have on its performance. As I've remarked before, the ubiquitous step of laying out an amplifier's circuit on a printed circuit board sometimes introduces a surprisingly large degradation in sound quality, even though the circuit is identical to that of the hard-wired "bird's-nest" prototype. I've often thought that it is the reduction of the circuit's physical realization from three to two dimensions that conspires to rob it of some of its sonic transparency, and I'm sure that is an experience shared by many readers! As the years have gone by I've periodically revisited this problem through such steps as investigating the choice of board laminate material and by being particularly careful about the relative positioning of the components on the board.

However I never really had a lot of motivation to solve the problem until last year when I was designing a gas sensor with a resolution of less than one part per billion. However you cut it that is a very small number and in ratio terms it must be equivalent to a dynamic range of some 180dB. So it was no surprise that some of the interactions; distortions and interference signals in the electronics proved to be quite troublesome. As part of this work I gained possession of some almost microscopic magnetic sensors which proved capable of measuring tiny varying magnetic fields and I was momentarily surprised to measure small ac currents around the various components used on the circuit board. I say momentarily because I quickly remembered good old Maxwell's equations which are every schoolchild's introduction to electromagnetism. As a result of his and Ampere's experiments we know that a current flowing through a conductor will generate a magnetic field around the conductor and that when a conductor cuts through a magnetic field a current will be generated in that conductor. All of which is a salutary reminder that electronics is largely a business of electromagnetic fields and currents rather than the more convenient shorthand of voltages which normally fill these pages.

And it is important to remember that the conductor doesn't need to be a magnet; it doesn't need to be a magnetic material, it just needs to be conducting an electrical current. If that current is an alternating current then the magnetic field will alternate. Put a second conductor next to the first and that alternating magnetic field will generate an alternating current in that second conductor and lo and behold we have created a transformer. The same process is also valid for components which are not simple wires; for example resistors. Typically a resistor comprises a rod coated with a resistive (or poorly conducting) coating between two connecting wires. Park a micro search coil next to the resistor and pass a signal through it and you will detect a matching magnetic field. Put a second resistor alongside the first and you will detect current flow through that second resistor. Depending upon the resistor type and value this current is likely to be very, very small but it is there. The same effect is there with some capacitors depending upon their construction and size. And it gets worse because not only can the current flow in one copper track induce current into an adjacent track but a track can induce current into a resistor and vice-versa. I'm sure that by now you can see where I am going with this. The strength of a magnetic field reduces with distance following an inverse square law or an inverse cube law depending upon how you view it; whatever it is, it reduces rapidly with distance. Thus a traditional "rats nest" prototype with components hung in the air is going to have few serious electromagnetic interactions or they are going to be very small. However once all the components are laid out neatly on a board with equally neat parallel tracks there must be lots of cross-coupling of signals albeit at very low levels.

Looking at a typical board layout and associated cable loom shows no end of first, second and higher order interactions which could be creating a layer of signal related noise. From this supposition I now had a theoretical explanation for my oft heard problem and, with the aid of some very careful measurements, a degree of evidence that there might be just enough low-level interactions to make my theory a practical possibility. At that stage my focus shifted to taking even more care with my board layouts and to shifting almost fully to surface mounted components because I reasoned, not totally accurately as it turned out, that the interactions might be fewer.

There matters rested for a few years until I was working on a switch-mode power supply running at over 300 kHz. I had a problem which I eventually traced to the ferrite transformer core saturating. The core was fully magnetised; couldn't adsorb any more flux and so the whole transformer process seized up. Shortly afterwards there was that familiar hot smell and the MOSFETs blew apart. Not wishing to delve into the charred mess for a while I took to pondering until I was struck by the thought that if a magnetic field is fully saturated then no additional field can be added or make any difference. So it then follows that if an amplifier were contained within a saturated magnetic field then the conductors; resistors; capacitors etc. could not create those small signal related magnetic fields (**figure 1**). This fundamental insight was soon demonstrated in the laboratory with the help of some hefty magnets and my sensors. But would this prove to be an interesting observation or would the effect be audible?

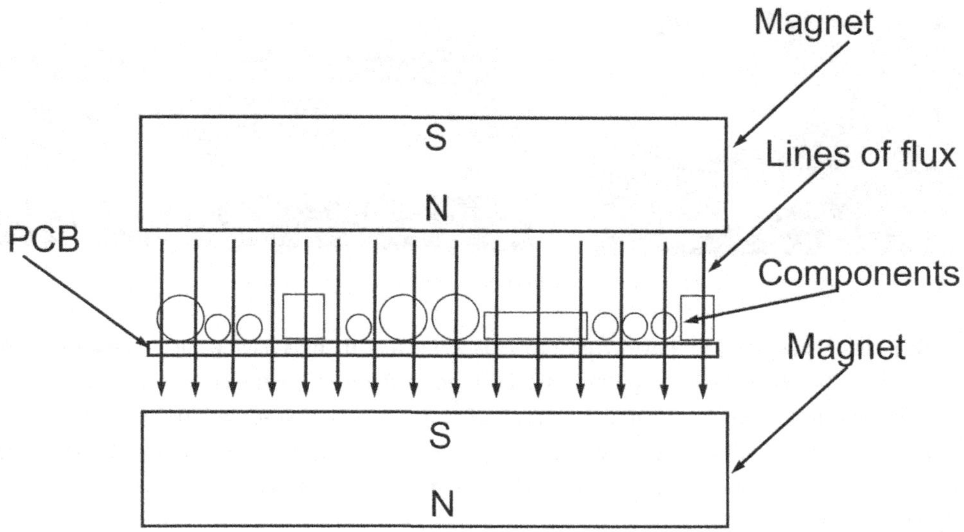

Figure 1 Concept for field saturating the unit.

For my initial experiments I took two sheets of MDF and machined out cavities to house some ferrite magnet cores recycled from old loudspeaker drive units. The two boards were placed above & below an amplifier with the North poles facing the South poles so creating a permanent static magnetic field through the electronics (**figure 2**). The audible effects were immediate as I put the boards in and out of position and repeatable until the flux started to induce some short-term residual magnetism into the magnetic parts of the amplifier. Subsequent experiments showed that the presence of solid copper conductors near each magnet assembly enhanced the effects and in due course, having found the optimum arrangements I filed some patents just in case. Having heard the effect of my crude boards I was soon visiting trusted friends in the industry and was heartened to find that I wasn't imagining things and then one UK electronics manufacturer asked if the technique could be built into an amplifier. Things now took a very interesting turn as I began positioning small assemblies (magnet + conductor) in only a few specific locations yet still demonstrated clear benefits.

Such work continues at quite a plodding pace because I really don't have the spare time for this project right now. Currently I have little commercial inclination to take things forward but I'm really interested in what others make of it all. So I had a batch of boards made and distributed some of them to about 50 enthusiasts; probably ten of whom could be described as "professional listeners" the rest being people who enjoy listening to music. Without exception all the listeners heard a difference with the positive comments running about 80:20 in favour. Those in favour felt that the sound was more open with the musical strands more clearly differentiated. Those against generally felt that the sound was more forceful and less musical. In fact the comments seemed quite like those made in analogue versus digital debates.

Figure 2 Initial test setup (left) and production-ready panels.

So dear readers what do we have? Well if this magnetic treatment is having such an audible effect then it must be doing something to the signal. If most listeners view the change positively then it must be doing something good to the signal. And what exactly is that something? As you have read I have my theories and they may be correct but I'm sure that with the enthusiastic readership of *Linear Audio* we'll soon have a few more theories and maybe one of us will get to the bottom of my interesting story.

Product Review: Smyth Research Inc. Realiser A8

Jan Didden and Stuart Yaniger

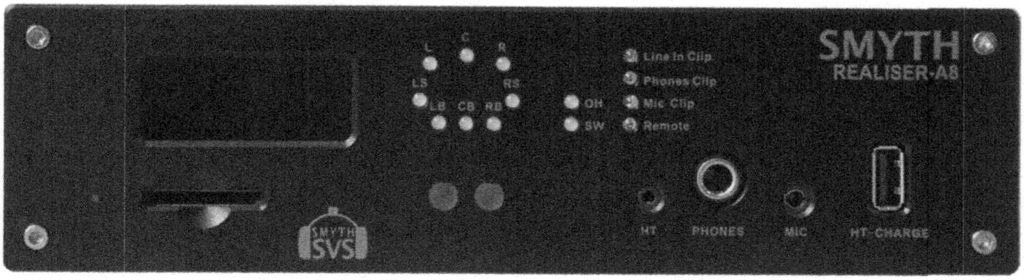

Introduction

One of us (JD) heard a demo of the A8 Realiser in Germany several years ago. The demo made quite an impression but due to other ongoing projects, follow-up was filed under 'when I next have some time to spare'. While preparing for the 2013 October series of audio events (Burning Amp, Rocky Mountain Audio Fest, Audio Engineering Society convention) the Realiser came back to the fore as an interesting piece of equipment to demonstrate and discuss at these shows. Those demos would also provide user opinions that could be integrated into a review. Smyth Research gracefully agreed to provide a unit with matching headphones several weeks ahead of the first event. This was important – the Realiser is a very unusual piece of equipment so getting familiar with it and its operational possibilities requires some time.

Background

Headphones are a convenient way to listen to music, but they have several disadvantages. One major problem is that the sound seems to be localized inside the listener's head, rather than spaced away from the listener (usually in front) as in live music or conventional stereo or multichannel. The problem seems to be exacerbated with multichannel sources. Over the years, this issue has been understood by considering the crosstalk that real acoustic sources have to the ears, including amplitude and timing, as well as frequency response due to diffraction and absorption by the head and pinnae. To solve this problem, engineers have developed a dazzling variety of signal-processing methods that are intended to provide a more natural sense of localization.

Interaural Timing Difference (ITD) and Interaural Level Difference (ILD) are known phenomena that help explain how we know something is to the left or right, but they cannot convey front/back or up/down, or range. The ITD and ILD will be identical at a particular azimuth angle whether in front or behind, above or below, near or far. It is the head-related transfer function (HRTF) that supplies the full angular localization information (azimuth and elevation). Range (distance to source) is determined by the brain from changes in the HRTF with head movement, and also to some extend using the ratio of direct to reverberant energy.

The earliest applications of these models to the issue of localization in headphones of which we are aware is that of Benjamin Bauer in the 1960s, who offered some analog methods of ameliorating the problem- Bauer's approach used some clever equalization and delay of a signal cross-fed from one channel to the other in an attempt to simulate the effect of ears in two different locations.

These analog implementations can improve the headphone "performers in your skull" problem, but in our experience, the improvement is rather limited. With the recent availability of cheap and fast signal processing in the digital domain, designers have been able to experiment with delay, EQ, and other methods with a flexibility that they have never had in the past when limited to opamps and wire. Thus, it isn't terribly surprising that software and firmware approaches have pretty much displaced analog approaches- the amazing power of computation and the ease of modification/upgrade are far too attractive to pass up.

What it does and how it does it

And this brings us to the Smyth Research Inc. Realiser A8, which incorporates the signal processing into firmware. The easiest way to describe the Realiser is probably as follows: 'a listening venue virtualiser'. The A8 provides, through its headphone replay, the perception that you are listening to the sound in your listening room, using your own audio system including speakers, with your own ears. Of course, to be able to do this, the A8 must 'know' about your sound system, your room and your ears. This information is gathered through a set of calibration and measurement runs. The A8 comes with a pair of miniature in-ear microphones (**photo 1**) which are fitted in the ear canal, and receive the sound at almost the same way your ears would receive it. It is not perfect, as the mics themselves are blocking the last fraction of an inch of the canal, but it is close. More on this later.

The calibration starts with the A8 playing some test sounds through the speakers to help the listener calibrate the replay level so that the signal to noise ratio at the in-ear mics is sufficient for the required measurement accuracy.

Next, sweeps are played through each speaker in turn. In a stereo system, the wearer of the mics is instructed to 'look left', 'look right' and 'look center'. This process allows the A8 to record the combination of speaker response, room acoustics and the user's Head Related Transfer Function (HRTF). The set of recorded data is saved as the Personal Room Impulse Response (PRIR). Finally, the user is instructed to place the headphone over the ears with the mics still in place, and test sweeps are

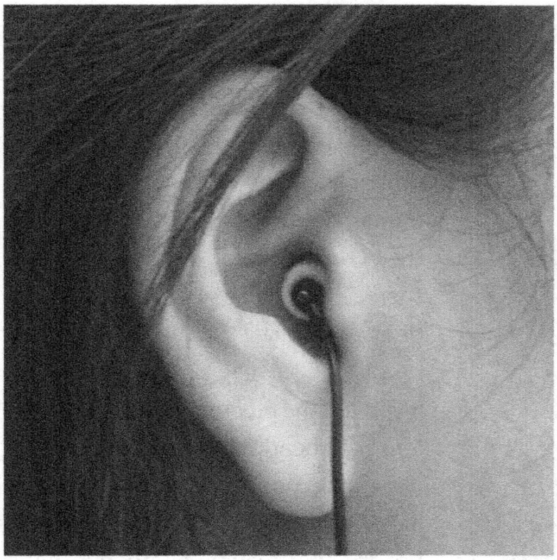

Photo 1 The miniature measurement mike sits on the outward side of the ear plug.

played through the headphones to record the transfer function from headphones to the ear. This latter set of data is saved as the Head Phone EQualization set (HPEQ). PRIR and HPEQ can be labeled to identify the user and/or the venue where they were recorded.

The HRTF contained in the PRIR allows the A8 to 'externalize' the perceived sound and to provide the illusion that one is not listening to headphones but rather to speakers at some specific distance and angle from the listener. HRTF refers to the way sounds are modified as they strike the upper torso, head, and especially the external ear (pinna) before the sounds enter the ear canal. These modifications are changes to the spectrum, but the brain does not hear the filter effect consciously, rather, it translates this into direction.
Because heads and pinnae vary widely, only a measurement through one's own pinnae can provide an emulation which is very nearly perfect, though one could listen to someone else's measurement for casual listening and there would be no ambiguity as to basic virtual speaker position. It just would not be precisely right, and the EQ would also be off since various heads sound brighter or duller to a foreign listener

Besides HRTF processing, there is a secondary mechanism by which the A8 helps the externalization, namely Head Tracking. A small IR transmitter is placed on the headphone headband on top of the listeners' head (**photo 2**), and a matching receiver is placed at a convenient place on the wall where the virtual speakers should be perceived. As the listener rotates his/her head, either consciously or subconsciously, the receiver detects this and causes the A8 to move the sound field in the opposite direction, thereby giving the illusion that the sound field stays in the same lateral position. This reinforces the externalization process, and the combined effect is a very convincing illusion that one is listening to his own speakers in his own room, complete with all the room influences that play a role.

Hardware Implementation

The Realiser is packaged in a cute little black box with an external "wall wart" supply. It has RCA analog inputs (unbalanced) and both RCA outputs and a ¼" jack for headphones and to provide a bypass to send the signal to the usual power-amp and speakers. The digital I/O is HDMI- no coaxial or XLR-type S/PDIF nor AES/EBU, which we would consider a major oversight- this required some major reconfiguring of SY's system, since his usual digital sources are run to the coaxial S/PDIF input of an

Photo 2 The head angle transmitter on top of the headphone band.

electronic crossover. If you're a multi-channel user or have HDMI-compatible digital sources, you're good to go all digital, but if not, you'll have to do all the I/O in the analog domain. Curiously, there is a Toslink fiber optic output-does anyone still use these?

If you use the recommended and excellent Stax headphones (which we did), the analog outputs of the A8 drive the headphone amp directly. Dynamic headphones may be plugged into the ¼" phone jack provided on the front panel.

The fun begins with the head tracking setup, which comprises a head-mounted IR transmitter and an optical angle-sensitive receiver. The receiver is mounted front and center of the listener, using a plastic clip which is big enough to clip on top of most relatively thin TV sets. The transmitter clips onto a bracket sized for the Stax headphones- other makes, be prepared to improvise. When mounted on the head, it gives the user a bit of a Teletubbies appearance. To avoid terminal embarrassment, we suggest the head tracker be used only in private or in the presence of consenting adults who promise not to tell.

Besides the Dork Factor, the main disadvantage of this implementation is that the optical receiver needs to be connected to the A8 black box via a long wire. That's always nice to have running down the middle of the living room floor. It's less of a problem for viewers of desktop or laptop video screens, but it also means that the head tracking function can only be used in a room where you can mount the receiver. This seems a curious implementation, given that tilt and angle sensing is now a commodity, being contained in most smartphones and game controllers. One could built an entire unit, including a wireless connection between the sensors and the A8 black box, to fit into the current head-mounted transmitter's form factor, and do it inexpensively. This would also allow "Realised" files to be listened to in portable situations. (On the other hand, the choice for this particular setup may have made perfect sense when the A8 was designed, five or six years ago, when integrated sensors were not generally available and rather expensive. Technology moves very fast these days).

Another key hardware piece is the previously mentioned pair of electret microphones, which are inserted into the user's ears, flush with the opening of the ear canals. These are used in two steps- the calibration of the room, performed with the headphones off, and the calibration of the headphones, performed (obviously) with the headphones on. Yes, the process looks as dorky as the head tracker, especially for the room calibration, where the head tracker needs to be mounted on a hair band (the cosmetic kind, not the rock kind!). But fortunately, this only needs to be done once per user.

Use and documentation

The documentation seems to have been written by an engineer rather than by a skilled technical writer. It is difficult to follow, does not emphasize the important points, presents information out of order, no "quick start" guide for most common uses... think of trying to learn Microsoft Word by reading the manual! Like with Word, there are so many features, settings, and options that one can easily get lost in the thicket, and there's no guide to extricating oneself.

The A8's manual can be a valuable reference once the basic setup process is understood, but it's tough going if you want to use it to learn how to do the basics of setup. The user should take advantage of the experience of Smyth's dealers and insist on a training session and an initial setup. The default setup provided with the A8 is, to be honest, lame in the extreme. We could hear that things were different processed versus unprocessed, but neither of us got an illusion of "out of the head" imaging. Once we struggled our way through the process, we were rewarded with a stunningly real illusion- but the learning curve was unnecessarily steep.
A few hundred dollars to a good technical writer could fix this issue completely.

How does it sound?

Before going 'on tour' as it were, with the A8, we did several measurement and listening sessions at SY's residence. After several attempts, mainly due to the unfamiliar terminology and structure of the user guide, we produced PRIR and HPEQ measurements for both our ears in his listening environment. Once that was done, the A8 produced extremely convincing and accurate reproduction of the sound stage on the headphones. The remote control allows switching between speakers and headphones and we found it very difficult to decide whether we were listening to the speakers or the headphones. This may sound, on the face of it, as an exaggerated statement, but it has to be experienced to be fully appreciated. At one point at an early morning testing session, while SY's wife was still asleep, with the headphones on and the A8 switched in, a musical selection started and JD reflexively ripped off the headphones and dived for the volume control, thinking that the loud sounds were coming from the speakers! This sort of complete fool-the-ear reaction is a rare and amazing thing – except with the A8!

Still, we are both very well aware of the usual bias and other traps involved in subjective listening and wanted to do some more testing with less involved individuals. This chance came when discussing the A8 with Siegfried Linkwitz, a known expert on speakers, room acoustics and all the factors involved in providing convincing music reproduction. Mr. Linkwitz kindly invited us to bring the A8 to his residence and compare its performance with his personal, highly optimized speaker system.

After performing the PRIR and HPEQ calibration runs we proceeded to listen to Linkwitz' favorite music. We used a remote control to switch the speakers on and off while Linkwitz took the headphones on and off, to allow an instantaneous comparison[1]. Linkwitz commented that, if we timed the switchover right, he could not hear any difference in sound stage between the speakers and the headphones. He was impressed.
There was however a slight timbral difference between the two situations. According to Smyth Research, this may be attributable to the fact that the mics measure at the exit of the ear canal ("blocked meatus") while ideally one would want to measure at the eardrum. There are obvious practical problems with doing that, and a kind of 'universal correction' is also not possible because people's ear canals are often very different.

However, the timbral difference can be compensated using the "Manual EQ" method described in the A8 manual. In this mode, the Realiser generates bands of pink noise, and for each band the listener compares headphones on / headphones off. For most bands the two will be the same, but when they are not, an adjustment can be made. In this way the timbral difference can be completely removed. It is not a difficult procedure, but it is tedious. One should ask whether any slight difference is relevant, that is, is it on the order of the same model speaker made this year instead of last year, or a slightly different position in the room, etc. According to Smyth Research, most pro users find the difference, if any, irrelevant. Mastering engineers sometimes use Manual EQ and they seem satisfied with the result. Manual EQ is an adjustment to the HPEQ file, not the PRIR.

It should be noted that the above described ear canal effect is subtle and not necessarily perceptible. A large difference suggests an error in measurement, perhaps poorly located mics; we found this out the hard way when we got a bizarrely unbalanced result from the first PRIR measurement performed using SY's ears. Following advice from the manual, we retried the measurement with the microphones correctly reinserted in his ear canals, which fixed that immediately.

Casual demos

As planned, we took the A8 to various audio events that took place in October 2013. Burning Amp is a gathering of audio diy aficionados in San Francisco; the Rocky Mountain Audio Fest is a three-day event featuring manufacturer listening rooms, a parts, equipment and records/CDs market place and technical presentations, and takes place in Denver, Colorado. Finally the Audio Engineering Society held its 135th convention in New York. At each event, we demonstrated the A8 to interested participants.

At these casual demos, it was not possible to record the PRIR and HPEQ data for each listener as there was no speaker system and listening room available. The system was played with the PRIR and HPEQ recorded at an earlier session. Therefore, it was not possible to compare the headphone rendering

[1] There is an automatic function in the A8 that switches between headphones and speakers when the headphone is tilted downwards, but we did not use this at the time.

with speaker rendering, but it was still possible to judge the believability of the externalization. The head tracking system was used in each case, to help the process. The remote has the capability of switching the processing on or off, so one can instantaneously compare the A8 externalization processing with the situation of just L and R headphone stereo with its familiar 'in the head' sound field. In all cases, listeners were very impressed with the results of externalization. Generally, when we switched to 'just headphones' in the middle of a listening session, listeners reacted strongly to the collapsing sound field and were often literally speechless. Many people asked where they could buy it and what it cost[2]. While such impulsive reactions do not necessarily lead to a purchase, they do illustrate how convincing the experience is.

Surround sound use

In all above cases, the A8 has been used in 2-channel stereo mode, as we both predominantly listen to stereo. However, the A8 is clearly targeted to surround users as well. One of us did some experiments in this mode, using a very clever function in the A8 to set it up for full surround using only a single speaker! Remember that to calibrate the system for the stereo setup in a room, the user has to alternatively 'look left', 'look right' and then the sweeps are played and the response in that look angle recorded as part of the overall PRIR.

This concept can be extended to 7.1 surround in several ways. One method is to use just a single speaker, and instruct the A8 to record the PRIR successively as follows. Just look forward (speaker at 0^0 angle) for the first sweep, then move the speaker to the $+30^0$ listening angle position and repeat the sweep, continuing in 30^0 increments until a full circle has been recorded. Thus the A8 records the responses obtained as if there were a full system of surround speakers in each of those positions. And you can even do it simpler: the 'ONE' procedure allows you to just leave *one* speaker in *one* fixed position, say front center, and instead move your head to each of the called out listening angles. There is a difference between the two procedures of course: when moving the one speaker to each position, the A8 does actually record the room response in each position. When leaving the speaker stationary but changing listening angle, you can record a kind of 'artificial' room response using the preferred position in your room. Which one gives the preferred rendering depends, among other factors, on the symmetry of the room, and can even provide better replay acoustics than would be possible in the actual room.
This method to synthesize a full surround system from a single speaker has one great advantage: you can use one very good, very expensive speaker instead of having to buy 7 or 8 of them! If you own a very high quality stereo speaker system, you can now synthesize it into an equally high quality surround system!

After setting up the system with one stationary speaker as described, we proceeded to listen to the surround version of the Avatar movie DVD. Again, as experienced with the stereo tests, the surround

[2] The retail price for the A8 Realiser was $ 2910 at the time of writing. While the A8 can be used with any headphone, a Stax SR-207 with matching SRM-2525 amplifier is recommended and can be provided by Smyth Research, and would add $ 760 to the total.

rendering was totally convincing. When watching the movie, the sounds were placed in an external 360^0 sound field and matched very precisely the way the characters in the movie looked towards the sound sources.

Wrap-Up

We generally find the physical pressure and sometimes 'sweatiness' of headphone ear pieces objectionable. But when listening to the Realiser, you are unconsciously more focused on an imaginary set of speakers. The result is that you tend to 'forget' that there is a set of headphones on your head, and the objectionable physical presence recedes to the back of your conscious focus.

We also find the effect of virtualization of speakers and sound field more 'natural' if you can see an actual set of speakers, even if they are not playing. We'd like to test that also with just speaker mock-ups visible, or maybe just pictures of speakers on the wall. Don't laugh – perception is all about make-believe! At any rate, this effect tends to diminish when you listen more often to the A8 and get used to hearing a speaker system without one being visible.

A common meme in fashion audio circles (especially in response to skepticism about things like "sound" of amplification or wires or demagnetization of CDs) is that, "Scientists and engineers know very little about how humans hear or the details of psychoacoustic processing in the brain." Although one could suggest in vain that people who say this spend some time reviewing the hundreds of books and thousands of papers written about this very subject, perhaps it would be faster to have them spend ten minutes with the remarkable Smyth Realiser A8. Designed using solid principles of engineering and well-established understanding of human hearing, with no magic or voodoo, its function is startlingly good.

Our only reservations are with some of the ergonomic features and documentation, not with the actual concept and execution of this remarkably convincing signal processing device. If you do a lot of headphone listening, or you want to 'take your system with you', as it were, consider an audition of the Smyth Realiser Inc. A8.

Further Reading:

Benjamin Bauer, "Stereophonic Earphones and Binaural Loudspeakers," JAES, April 1961.

Benjamin Bauer, "Improving Headphone Listening Comfort," JAES, October 1965.

Louis A. Abbagnaro, , Benjamin Bauer, and Emil Torik, "Measurements of Diffraction and Interaural Delay of a Progressive Sound Wave Caused by the Human Head," JASA September 1975.

Frederic Wightman and Doris Kistler, "A Model of Head-Related Transfer Functions Based on Principal Components Analysis and Minimum-Phase Reconstruction," JASA, March 1992.

Frederic Wightman and Doris Kistler, "Headphone Simulation of Free-Field Listening" JASA, February 1989.

Stephen Smyth, Personalized headphone virtualization, US Patent 7,936,887.

JM Loomis , C Hebert , and JG Cicinelli, "Active localization of virtual sounds," JASA, October 1990.

Book review – Building Valve Amplifiers

Marcel van de Gevel

Building Valve Amplifiers – Morgan Jones. 2nd edition, Newness, February 2014. ISBN 978-0080966380. Amazon printed edition $ 42, Kindle $ 36.

Although its title indicates that Morgan Jones' book Building Valve Amplifiers is a book about building valve amplifiers, much of its contents is equally applicable to other electronic circuits. That includes transistorised circuits, as Jones often uses transistors in supporting circuitry and sometimes even in the signal path. Hence, the book is really about constructing electronics at home, with emphasis on chassis-mounted valve amplifiers. It is explicitly not a book about design issues, but about constructing and testing already designed circuits.

Despite Jones' preference for metric units, the book is very British: not just the language and his humour, also the references to fuses in mains cords, mains outlets with on/off switches and the +10 %/-6 % tolerance on the European 230 V mains (correct in the UK, but it is +6 %/-10 % in most of the European mainland for compatibility with 220 V equipment). Regarding the language, there is a translation table included in the appendix for American readers.

If you think that 651 pages (excluding preface) is a lot for a book about building already designed electronics, you are probably right. Nevertheless, there is no undue redundancy in the book. The high page count is simply due to the thoroughness with which Jones deals with the subject.

The book is divided in the sections Construction, Testing and Examples. The first section consists of three chapters, about layout, metalwork and wiring. The chapter about layout deals with issues such as magnetic coupling and cooling and already included a couple of eye-openers for me.

The chapter about metalwork explains how to make a professional-looking chassis and what kind of reasonably affordable tools you need to do this. As a non-native speaker of English, I am happy that Jones included photographs of the tools he discusses; for most of these things I don't even know what they are called in my own language, and I certainly don't know it in English. Anyone who has ever seen a piece of electronic equipment built by me knows that I'm not the right person to comment on the contents of this chapter; I haven't got most of the discussed tools and my chassis and enclosures typically look anything but professional[1].

The wiring chapter deals with soldering, with how to obtain nice, tightly twisted heater wiring and with earthing/grounding. (Apparently earthing is the usual term in the UK and grounding in the US; I'll use earthing for safety earth and grounding for signal ground.) Although I have soldered since primary school, even the part about soldering still contains some information that is new to me.

The part about earthing shows that Jones and I have different preferences for what to do with the chassis/metal enclosure of an amplifier with unbalanced inputs. I usually regard the chassis as a ground plane for signal ground, take for granted that I can't connect it to safety earth (at least not on more than one piece of equipment) because of hum loops and try to make the equipment double isolated. In the Netherlands, an earthed mains plug only provides false security anyway, because most houses only have earthed mains outlets in the kitchen and bathroom and because an earthed plug fits in a mains outlet without earth. Jones always firmly connects the chassis to safety earth and takes for granted that he may have to keep the signal ground separated from the chassis to avoid loops. As a result, he uses star grounds or ground bars that result in relatively long ground connections and relatively large internal loop areas, which is bad for EMC, or he resorts to using a PCB with a ground plane. In any case, Jones clearly describes the various methods and their main advantages and disadvantages, enabling the reader to make up his or her own mind.

The one thing that isn't clear to me, though, is how Jones can separate the signal ground from the chassis and provide connections to outside equipment without violating the rules for electrical safety of class I equipment. A cinch connector that is insulated from the chassis would be a piece of metal that you can touch, that is connected to the internal circuitry and that has no safety earth. Maybe you can circumvent this by, for example, using female DIN connectors, but as soon as you plug in a DIN cable, you have the same issue with the DIN-plug at the other end of the cable.

None of these issues play a role when you use AES48-compliant balanced connections instead of unbalanced connections. Jones' discussion of balanced connections is AES48-compliant, as it should be. The section with examples at the end of the book actually only contains AES48-compliant balanced examples.

[1] But they sound professional. See as an example: A tube-based phono preamplifier, Marcel van de Gevel, Linear Audio Vol 4 – Ed.

The second section of the book is about measuring, either to find faults or to determine the performance of the equipment. This section again has the usual thoroughness, although the discussion in chapter 4 about combining uncorrelated measurement uncertainties is somewhat oversimplified. It doesn't consider the probability distributions of the various uncertainties and the resulting probability distribution and probabilities of excess for the combined result. Then again, if it did, the book would turn into an introduction into statistics rather than a book about how to build a valve amplifier.

Chapter 4 introduces the main test equipment, for example analogue and digital oscilloscopes. It is a pity that Jones doesn't mention analogue variable-persistence storage oscilloscopes, as these already had some of the features that he likes most in digital oscilloscopes decades before anyone even considered digitising oscilloscopes. What I also missed is a warning that the feature of digital oscilloscopes to dump data to a file is only useable for a couple of years, namely until all computers that support the file format or the data carrier have been moved to the museum.

The part about how to accurately model a transformer is very clear and thorough, but would seem more at its place in a book about circuit design. I don't know why Jones decided to include it in this book, maybe because valve amplifiers are often built with other transformers than the designer intended?

Chapter 5 deals with electrical safety, functional testing and faultfinding. Chapter 6 is about performance testing.

A weak point in chapter 6 is the paragraph about weighted noise measurements. Jones doesn't like A-weighting because it results in higher signal-to-noise ratios than unweighted RMS measurements, thereby making marketeers happy. He also doesn't like weighted measurements according to ITU-R 468 (formerly CCIR-468) because they give subjectively more or less correct results for impulsive noise as found on analogue music lines.

This doesn't make much sense to me; despite its imperfections, noise weighting gives measured results that correspond better with the subjective impression of a human listener. If the aim of a measurement is to get subjectively meaningful results and to make marketeers unhappy, ITU-R 468 is definitely the best of the standard audio noise measurement techniques, because due to its quasi-peak detection, it invariably produces lower values for signal-to-noise ratios than unweighted 20 Hz to 20 kHz RMS measurements. Valves usually have different optimal bias currents for minimal white noise and minimal 1/f noise, which makes it especially important to take at least a rough approximation of the frequency-dependence of the human auditory system into account.

The part about RIAA stage noise shows that Jones is well aware of the importance of input current noise and the frequency-dependence of the source impedance. Unfortunately, he falls in the trap of thinking that JFET and valve RIAA stages have negligible current noise. Moving-magnet RIAA stages with or without valves usually have a substantial noise current source right at their input, namely the 47 kohm termination resistor.

Chapter 7 is the examples chapter, showing some well-designed and beautifully constructed hybrid amplifiers.

Summarising, this is a book for anyone interested in building electronic circuits at home, both those who are just starting and those with considerable experience. It is incredible how much useful information Morgan Jones has collected and put into this book. On top of that, Jones writes in a very clear and pleasant style. Highly recommended.

Made in United States
North Haven, CT
10 November 2026

22096537R00115